BLACK HOLES
WORMHOLES

& TIME MACHINES

About the Author

Jim Al-Khalili was born in 1962 and works as a theoretical physicist at the University of Surrey in Guildford. He is a pioneering popularizer of science and is dedicated to conveying the wonder of science and to demystifying its frontiers for the general public. He is an active member of the *Public Awareness of Nuclear Science* European committee. His current research is into the properties of new types of atomic nuclei containing *neutron halos*. He obtained his PhD in theoretical nuclear physics from Surrey in 1989 and, after two years at University College London, returned to Surrey as a Research Fellow before being appointed lecturer in 1992. He has since taught quantum physics, relativity theory, mathematics and nuclear physics to Surrey undergraduates. He is married with two young children and lives in Portsmouth in Hampshire.

BLACK HOLES
WORMHOLES
& TIME MACHINES

JIM AL-KHALILI
University of Surrey

Published in 1999 by
Taylor & Francis Group
270 Madison Avenue
New York, NY 10016

© 1999 by Taylor & Francis Group, LLC

No claim to original U.S. Government works
Printed in the United States of America on acid-free paper
15 14 13 12 11 10 9

International Standard Book Number-10: 0-7503-0560-6 (Softcover)

Library of Congress Cataloging-in-Publication Data

Catalog record is available from the Library of Congress

Taylor & Francis Group
is the Academic Division of T&F Informa plc.

Visit the Taylor & Francis Web site at
http://www.taylorandfrancis.com

To Julie, David and Kate

CONTENTS

TIME

TIME MACHINES

PREFACE

Over the past few years there has been an explosion in the number of books and television programmes popularizing current scientific ideas and theories and making them accessible to a wider audience. So is there any need for this, yet another book on a subject that has received more attention than most: the nature of space and time and the origin of our Universe? The other day, I was looking through the web site of a large Internet book club. Under the category of science and nature, I searched for all books with the word 'time' in their title. I found 29! Of course, Stephen Hawking's *Brief History of Time* is the best known of these, but there were many others with titles like *About Time*, *The Birth of Time*, *The Edge of Time*, *The River of Time* and so on. It seems that questioning the nature of time at a fundamental level is the 'in' topic at the moment. What was most surprising for me was to see that many of those 29 titles had been published *since* I began writing this book.

Established science writers such as Paul Davies, John Gribbin and Richard Dawkins were an inspiration to me as an undergraduate in the mid-1980s. But they were preaching to the converted. At best, they were aimed at the 'intelligent layperson', whoever that is supposed to be. My ambition has therefore been to write a book at a more basic level, which would explain some of the ideas and theories of modern physics for *anyone* to understand, provided of course that they are interested enough to pick up such a book in the first place. I have also tried to make it a little more fun, aiming (probably without much success) for a sort of Stephen Hawking-meets-Terry Pratchett.

Many scientists would argue that difficult subjects like Einstein's theories of relativity can only be 'dumbed down' so much before reaching a level where the explanations are no longer correct. I hate that term: dumbed down. It sounds so patronizing. And while it is flattering to be considered by society to be more intelligent than everyone else, scientists are just people who have spent many years being trained to understand the relevant jargon, abstract concepts and mathematical formulae. The hard part is to translate these into words and ideas that someone without their training can appreciate.

Because of the way this book developed it has been written with a teenage audience in mind. However, it is aimed at anyone who finds its title fascinating or intriguing. It does not matter if you have not picked up a science book since you were fifteen.

So how *did* this book come about? Well, about three years ago the then head of my physics department at the University of Surrey, Bill Gelletly, suggested that I should give, as one of a series of lectures to first year undergraduates covering a range of general interest topics in modern physics, a lecture on 'wormholes'. Such a topic is certainly not part of a traditional undergraduate degree course in physics. In fact, fans of the TV series *Star Trek: Deep Space Nine* are probably better informed about wormholes than your average physicist. Anyway, I thought it would be fun, and proceeded to do some background reading in preparation for the lecture. On the day, I was surprised to find in the audience many students not on the course, as well as postdoctoral researchers and members of staff. There seemed to be something magical about the title.

Each year, my department sends out a list of speakers, from among its academic staff, and lecture titles to local schools and colleges. This is mainly as publicity for the department in the hope that these lectures might play a part in our recruitment drive to attract new students. I offered my 'wormholes' talk as one of these. Such was its success, I was asked by the Institute of Physics whether I would be the 1998 Schools Lecturer. This involved the substantially greater commitment of having to travel around the country giving the lecture to 14–16 year-olds, with audiences of several hundred at a time. And, having put a significant amount of preparation into this performance, I found that I had accumulated far too much fascinating material to cram into a one hour lecture and decided to put it all down in a book.

I have tried as much as possible to be up to date. In fact, when the manuscript came back to me from the publishers for final corrections and changes, I had to completely revise the chapter on cosmology. Due to recent astronomical discoveries, many of the ideas about the size and shape of the Universe had changed during the few short months since I had written that chapter.

Jim Al-Khalili
Portsmouth, England, July 1999

ACKNOWLEDGMENTS

Looking back over the second half of 1998 when the bulk of the book was written, I realise that I owe my greatest debt of gratitude to my wife Julie and my children David and Kate for putting up with me. Since I could not allow my writing to interfere with my full-time research work, it had to be done at home during the evenings and the weekends. I am also indebted to the following friends, family and colleagues for kindly reading through the manuscript and making so many constructive comments and suggestions: Julie Al-Khalili, Reya Al-Khalili, Richard Wilson, Johnjoe MacFadden, Greg Knowles, Simon Doran, James Christley, Ray Mackintosh, John Miller and James Curry. I am sure that errors still remain, for which I hold sole responsibility. I must also thank Bill Gelletly for making the suggestion that got the whole project started, Kate Jones for some constructive lunch-time discussions on entropy, Youcef Nedjadi for clarifying some aspects of free will, Matt Visser for filling me in on some of the latest ideas about wormholes, Brian Stedeford for useful insights into the work of Lewis Carroll, Phil Palmer for clarifying a number of points in cosmology, James Malone for kindly providing the computer generated image of a wormhole for the book cover and finally my commissioning editor at Institute of Physics Publishing, Michael Taylor, for all his help and support.

INTRODUCTION

> *"The time has come" the Walrus said,*
> *"to talk of many things"* . . .
> Lewis Carroll, *Through the Looking Glass*

> . . . *of atoms, stars and galaxies,*
> *and what a black hole means;*
> *and whether Einstein's space can bend*
> *enough for time machines.*

This book is meant for all those people—which is pretty much everyone I know—who are curious about such exotic sounding concepts as black holes, space warps, the Big Bang, time travel and parallel universes. In writing the book I have asked myself whether complete non-experts can learn a little about some of the ideas of modern physics without feeling the urge to check that their IQ is up to the task before embarking.

The subject matter of the book has been covered elsewhere at many different levels. At the very top is the advanced text or monograph for the practitioner in the field. This is the sorcerer's spell book, decipherable only by the privileged few. Then comes the textbook aimed at the university physics student. It too contains some spells, but nothing very powerful. Below that comes the top end of the popular science market. Such books are aimed at the non-scientist in that they contain little or no mathematics. However, they appeal only to those who are either (a) other scientists or (b) fans of such books already, who have invariably read similar books on the subject.

So, when writing this book I have made every effort to cut out as much scientific jargon as possible. Popular science writers are, on the whole, becoming highly adept these days at explaining complex concepts using everyday words. But every now and then we will let slip a 'Jargonese' word which to us is so obvious we forget that it does not carry the same meaning for everyone.

Short or long ten minutes?

One summer, when I was about ten or eleven, I became fascinated with the concept of time. Where did it come from? Did we invent it or has it always been around? Does the future already exist somewhere? Is the past still being acted out? Deep questions for a kid. But, before you mistake me for a child prodigy, let me share with you what my idea of time travel was. I knew that on the other side of the world, somewhere in the middle of the Pacific Ocean, was an invisible line running from the North Pole to the South Pole which divided the world into today and yesterday! If a ship were anchored across this line then on one end of the ship it could be 9.00 on Tuesday morning and at the other end, still 10.00 on Monday morning. Surely this was a clear example of time travel, just by walking a few yards along the deck!

OK, I knew there was something fishy going on and I remember one evening my father explaining to me that time zones around the world are only man's invention. For instance if it is decreed that at midnight in New York it is already 5.00 am in London, this is just our way of making sure that, as the Earth spins, and different countries face towards the sun, the hours of daylight are roughly the same for everyone, if not at the same time. I followed all of this, sort of, but felt disappointed. Surely there was more to the concept of 'time' than that, something more mysterious. I had this theory about time flowing at different rates depending on my mood. Clocks definitely slowed down towards the end of school lessons and, as my birthday approached, the weeks and days almost ground to a halt.

Now it is the turn of my own children to come to these conclusions. If I tell them they have ten more minutes before they have to put their toys away, they are quite serious when they ask whether it is a short, medium or long ten minutes. Anyway, who can argue against the simple observation that, for a child, time goes by very slowly. One year is an extremely long time for five-year-olds since it makes up a fifth of their life, but the older we get the faster the years seem to flash by: can you believe it is Christmas again already!? or: has it really been three years since I was last here? and so on.

Deep down we feel we *know* that time flows at a steady rate. When asked how fast time flows the scientists' usual glib response is to say that it is at a rate of one second per second. In our culture we believe that, no matter how subjective we feel about the passage of time, there is a cosmic clock that marks off the seconds, minutes, hours, days and years everywhere in the Universe relentlessly and inexorably and there is nothing we can do to change it.

Or is there? Does such cosmic time really exist anyway? Modern physics has shown that it doesn't. Don't worry, there is very strong evidence to support this. In fact, before I go any further, try this out for size: we are certain that *time travel to the future is possible*. Scientists have successfully carried out many experiments that have tested this and proven it beyond any doubt. If you are in any doubt about this amazing, maybe even startling, piece of information then this is not due to any *X-Files*-type government cover-up but rather because you have not done a course in special relativity. All will be revealed, I hope, in this book.

Common sense

It is probably fair to say that most people are not exactly on best buddy terms with Einstein's theories of relativity (yes, there are two of them). So I am never surprised by the response I get when I tell my non-scientist friends that nothing can go faster than light. "How do you know?" they say. "Just because scientists haven't found anything yet that can go faster than light doesn't mean that you won't one day have to eat your words. You should be more open minded to other possibilities that just may not have occurred to you. Imagine showing a television to an isolated tribe in the deepest Amazon which has never seen one before," and so on. I am not in the least bothered by this response because it is exactly the attitude I would like the reader of this book to have. Namely, being open minded and having the ability to accept a new worldview even if it flies against everything you thought you were sure about, or what you would call simple common sense.

Albert Einstein was once quoted as saying that common sense is just the prejudices we acquire by the age of eighteen. So, for the Amazonian tribe which has never seen a television before, it would go against their common sense that such a box could speak to them and show them a whole world inside it. (OK, I am assuming that they have electricity there and a power point!) But I am sure you would agree that after we had spent enough time with this tribe explaining about radio waves and modern electronics and all the other things that go into making a television work, then they would grudgingly have to adjust their worldview so that this new information no longer went against their common sense.

At the beginning of the twentieth century, several new scientific theories were developed and proven to be, so far anyway, correct. Between them they are responsible for almost the whole of modern science and technology. The fact that we have digital watches, computers, televisions, microwaves, CD players and just about every other modern appliance is testimony to the fact that these theories are, if not the whole story, pretty much true in the way they describe the world around us. The theories in question are relativity and quantum mechanics. I should explain that a successful theory is one which can predict what would happen under certain circumstances: If I do *this* then according to my theory *that* will happen. If I carry out an experiment and find that the theory's predictions were correct then this is evidence in support of the theory. But a theory is not the same as a law.

The law of gravity says that all objects in the Universe are attracted to each other by a force that depends on how massive they are and how far apart they are. This is not open to doubt, and while we know that it needs to be modified when we are dealing with extremely massive objects like black holes, we trust it completely when it comes to describing the way falling objects behave on Earth. However, a theory is only good as long as a better one doesn't come along and disprove it. We can never prove a theory, only disprove it, and a successful theory is one that stands the test of time. Contrary to the view of many non-scientists, most scientists would like nothing better than to prove a scientific theory wrong, the more respectable the better. So, since theories such as quantum mechanics and Einstein's relativity have lasted for most of this century despite the constant efforts of physicists to prove them wrong or at least find loopholes and weaknesses, we have to admit that they are probably right, or at least on the right track.

Back to the future

Sorry, I am straying from the story. I should get back to the interesting stuff about time travel being possible. Later in the book I will explain in more depth what relativity theory is about. In the meantime, here is an example of what relativity has taught us. If you were to travel in a rocket that could go so fast it approached the speed of light, and you zipped around the Galaxy for, say, four years, then upon returning home to Earth you would be in for a bit of a shock. If your on-board calendar says you left in January 2000 and returned in January 2004, then depending on your exact speed and how twisted your path was through the stars, you might find that according to Earth the year is 2040 and everyone on Earth has aged forty years! They would be equally shocked to see how young you still looked considering how long you had, according to them, been away.

So your rocket clock, travelling at very high speed, had measured four years while all Earthbound clocks had counted off forty years. How can this be? Can time really slow down inside your rocket due to its high speed? If so, this means that, for all intents and purposes, you will have leapt thirty six years into the future!

Although I will come back to this later, the idea of time slowing down when you travel at high speeds is something that has actually been checked and confirmed many times in different experiments to extremely high degrees of accuracy. For example, scientists have synchronized two high precision atomic clocks, then placed one of them on a jet aircraft and the other in a laboratory on Earth. After the jet had returned, the two clocks were checked again. It was found that the travelling clock was a tiny fraction of a second behind its stay-at-home partner. Despite the modest speed of a thousand kilometres per hour at which the jet would have been flying compared with the speed of light (a further million times faster), the small, if unimpressive, difference between the readings of the two clocks is real. The clocks are so accurate that we do not doubt their readings or the conclusions we draw from them.

Readers who know something about relativity theory may wish to argue at this point that the above example is not as straightforward as I have made it sound. This is true, but the subtleties of what is known as the clocks paradox will have to wait until I discuss special relativity in Chapter 6. For now it is sufficient to keep the discussion at the level of the simple, but perfectly correct, statement that high speed motion allows time travel to the future.

How about time travel to the past? In many ways this is even more fascinating. But it turns out that it is also much more difficult. It might come as a surprise to you that travelling forward in time is easier than back in time. If anything, you might think that the notion of travelling into the future is the more ridiculous. The past may well be inaccessible, but at least it is out there; it has happened. The future on the other hand, has yet to happen. How can you travel to a time that has not happened yet?

Even worse, if you believe that you have some control over your destiny then there should be an infinite number of versions of the future. So what governs which version you would travel to? Of course, getting to the future by high speed space travel does not require the future to be already out there waiting for you. What it means is that you move out of everyone else's time frame and into one in which time moves more slowly. While you are in this state, time outside is ticking by more quickly and the future is unfolding at high speed. When you rejoin your original time frame you will have reached the future more quickly than everyone else. It is a bit like waking from a coma after a few years and thinking that you have only been away for a few hours. The difference there of course is that you will get a shock the first time you look in a mirror and see how much you have aged, whereas in the case of high speed travel your body clock and everything else in the rocket really is in a different time frame. What is really strange is that you don't notice anything different while you are moving at this speed. To you, time is going by at its normal rate on board the rocket and if you were able to look out of the window you would, paradoxically, see time *outside* going by more slowly!

There is a downside to this, however. Once you get to the future, you are stuck there and cannot return to the present you left behind. The date on which you left in your rocket is now in your past and time travel to the past is a bit of a problem. But calling it a problem is not the same as stating that it is impossible.

Meeting yourself

There are so many mind-boggling examples of how ridiculous things would be if time travel to the past were possible that I could fill this entire book with them. For example, what if time travel to the past were possible and you decided to visit your younger self at a time just before you were about to invest your life savings in a business venture which you know will fail. If you succeed in convincing your younger self not to go through with it, then presumably your life would have been different. By the time you reach the age at which you went back in time to advise yourself against the decision, there would be no need to do so since you never made the investment. So you don't go back. But at the same time you must have a memory of not investing the money because you were talked out of it by an older you who had visited you from the future. You now live in a world in which you made the decision *not to invest*. Was this because you met your older self who advised you against it? If so, how could you ever have become that person who felt the need to go back in time to warn you against something you didn't end up doing anyway?

If you are totally confused by what you have just read, don't worry, you are supposed to be. That is the whole point of a paradox. Here is what appears to be, at first glance anyway, a possible solution. If you do go back in time to warn yourself against doing something, then two things are true. Firstly, the fact that you are going back to the past to stop something that has already happened means that you must fail in that attempt because it *did* happen. There is, after all, just one version of history. Secondly, you should remember a time in your past when you were visited by an older you and you know that it had been a futile attempt and therefore know that it isn't worth trying. This is where this explanation breaks down. If you know it's no good going back to warn yourself and decide not to, then who did? You must go back in time because you remember meeting your older self who tried to convince you not to go into the venture. Somehow this means you have no freedom to choose your actions. So, what happens? Does some Time Lord appear and force you into the time machine warning you of the dire consequences to the very fabric of spacetime if you don't?

Despite such problems, you may be interested to know that time travel to the past was found to be allowed by Einstein's general theory of relativity, a discovery that was made half a century ago. And since general relativity is currently our best theory about the nature of time, we have to take its predictions seriously until we can find a good reason, possibly based on a deeper understanding of the theory, for why they might be ruled out. You may therefore be wondering why no one has so far been able to construct a time machine? In this book I explain why, touching on a few of the most fascinating topics in physics along the way.

Some of the things that we have discovered about our Universe are so amazing and incredible that I hope you will feel cheated that you hadn't known about them until now. That is what I would want you to get from this book; to share that feeling of wonder I have about the cosmos. That, and to give you some hard scientific ammunition with which to impress your dinner party friends when the time travel discussion gets going.

SPACE

1

THE 4TH DIMENSION

To do with shapes

Geometry is the branch of mathematics concerned with the properties and relations of points, lines, surfaces and solids. The majority of people probably don't look back at the geometry they learnt at school: the area of a circle, the lengths of the sides of a right-angled triangle, the volumes of cubes and cylinders, not forgetting those reliable tools of the trade, the compass and protractor, with nostalgic fondness. I therefore hope that you are not too put off by a chapter devoted to geometry.

In the spirit of this book's crusade against the scientific language of Jargonese, I will redefine the meaning of geometry by saying that it has *to do with shapes*. Let us examine what we mean by shapes in the most general sense. Look at the letter 'S'. Its *shape* is due to a single curved line. A splash of paint on a canvas also has a shape, but this is no longer that of a line but an area. Solid objects have shapes too. Cubes, spheres, people, cars all have geometric shapes called volumes.

The property that is different in the above three cases—the line, the surface and the volume—is the number of dimensions required to define them. A line is said to be one-dimensional, or 1D for short, an area is two-dimensional, or 2D, and a volume is 3D.

Is there some reason why I could not go on to higher dimensions? What is so special about the number three that we have to stop there? The answer, of course, is that we live in a universe which has three dimensions of space; we have the

3

freedom to move forward/backwards, left/right and up/down, but it is impossible for us to point in a new direction which is at right angles to the other three. In mathematics these three directions in which we are free to move are called mutually perpendicular, which is the mathematicians' way of saying 'at right angles to each other'.

All solid objects around us are 3D. The book you are reading has a certain height, width and thickness (all three quantities being lengths measured in directions at right angles to each other). Together, these three numbers define the book's dimensions. In fact, if you multiply the numbers together you obtain its volume. This is not so obvious for all solid objects. A sphere, for instance, needs only one number to define its size: its radius. But it is still three-dimensional because it is a solid object embedded in 3D space.

We see around us shapes that are either one-, two- or three-dimensional, never four-dimensional because such objects cannot be accommodated in our three-dimensional space. In fact, we cannot even *imagine* what a four-dimensional shape would look like. To imagine something means building a mental model of it in our brains which can only cope with up to three dimensions. We would, quite literally, not be able to get our heads round a 4D shape.

To many people, 'one-dimensional' means 'in one direction'. Adding another dimension to something means allowing it to move in a new direction. True enough, but, you might ask, how about that letter 'S'? When writing an 'S' your pen traces curves in different directions. How can the final shape still be 1D? Imagine a dot called Fred that lives on a straight line (figure 1.1). Fred is unable to move off the line and is restricted to movement up or down it. We say that his motion is one-dimensional. In fact, since the line is his entire universe, we say that Fred lives in a 1D universe. But what if his universe were the letter 'S'? How many dimensions would he be living in now? The answer is still one. He is still restricted to moving up or down the line. Granted, his life may be more interesting now that he has a few bends to tackle, but curving a shape does not increase its number of dimensions.

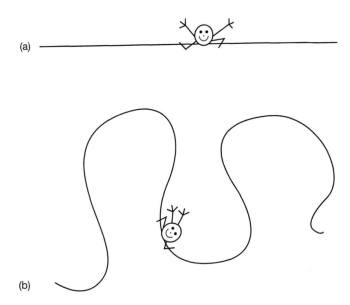

(a)

(b)

Figure 1.1. Fred the dot living in his one-dimensional universe that is (a) flat and (b) curved.

(By the way, since Fred himself is only a dot, or 'point' to give him his mathematical definition, he is thus a zero-dimensional being.)

Another way of talking about the dimensions of a space is by seeing how many numbers, called co-ordinates, we need to locate a certain position within that space. The following example, which I remember reading years ago but cannot remember where, is still the clearest one I know. Imagine you are on a barge going down a canal. Given some reference point, say that village you passed earlier, you need just one number: the distance you have travelled from the village, to define your position. If you then decide to stop for lunch you can phone a friend and inform them that you are, say, six miles upstream from the village. It doesn't matter how twisted the canal is, those six miles are the distance you travelled and not 'as the crow flies'. So we say that the barge is restricted to motion in one dimension even though it is not strictly in a straight line.

What if you are on a ship on the ocean? You now require two numbers (co-ordinates) to locate your position. These will be the latitude and longitude with respect to some reference point, say the nearest port or internationally fixed co-ordinates. The ship therefore moves in two dimensions.

For a submarine, on the other hand, you need three numbers. In addition to latitude and longitude you must also specify a length in the third dimension, its depth. And so we say that the submarine is free to move in three-dimensional space.

What is space?

During staff meetings in the Physics Department at Surrey University where I work, there is always an item on the agenda called 'Space'. This is where the different research groups argue over office space for research students and visiting researchers who need a desk for a few weeks and over laboratory space for their experiments. But when the head of department reaches that point in the meeting and says something like "And now we come to Space—", somebody usually mumbles "—the final frontier". And you thought physicists didn't have a sense of humour!

We all think we know what 'space' means, whether it is space in the sense of 'there is an empty space over in that corner' or 'not enough space to swing a cat', or space in the sense of 'outer space' of the final frontier variety. When forced to think about it we would regard space merely as somewhere to put things. Space in itself is not a substance. This much we would all agree on. But in that case, can space exist when it doesn't contain any matter? Think of an empty box. Even if we pump out all the molecules of air it contains so that there really is *nothing* inside the box, we would still be happy with the concept that the space continues to exist. The space refers only to the volume of the box.

It is less intuitive when space has no boundaries. The space inside the box only exists, we think, thanks to the existence of the box itself. What if we remove the lid and walls of the box? Does the space still exist? Of course it does. But it is now a region of space that is part of a larger region inside the room we are in. Let us try something a bit grander: Our Universe is basically a very large (maybe infinite) volume of space containing matter (galaxies, stars, nebulae, planets, etc). What if the Universe were completely empty and contained no matter at all? Would it still be there? The answer is yes, since space does not need to contain matter in order to exist. At this point, the discussion could easily nosedive—since I am doing all the discussing, and I know what I am like when I get going—into a highly technical and obscure (yet much debated) subject known as Mach's principle. This states that space, or at least distances and directions within it, is meaningless when it does not contain any matter. In addition to this, Einstein has shown in his theories of relativity that space, like time, is, well, relative. However, I do not want to get too heavy at this early stage of the book and will assume that although space is not a substance, it must nevertheless *be* something!

But if space is not a substance, can we interact with it? Can matter affect it in some way? It turns out that matter can indeed affect space itself: it can bend it! Once you appreciate this fact, you should never again be impressed with claims of cutlery bending by the powers of the mind (a cheap and rather pointless conjuring trick).

In the next chapter, I will be asking you to imagine bending 3D space[1]. That's OK, you might think, I can easily bend a 3D object such as this book. Well, it's not as simple as that. You see I don't mean 3D objects being bent *within* 3D space but rather bending 3D space itself.

Think about the curvature of the 1D line to form the letter 'S'. We need a 2D sheet of paper to write the 'S' on. We say that the 1D shape is imbedded in the higher dimension. Similarly, bending a sheet of paper requires the use of our 3D space if we want to visualize it. It follows that to appreciate what bending 3D space

[1] To be more accurate, whenever I discuss the bending of 3D space I should really say the bending of 4D 'spacetime'. This is what Einstein's theory of relativity says we have to call the combination of the three dimensions of space with the one of time. However, for the time being I will leave the discussion of how space and time get mixed up till later on in the book.

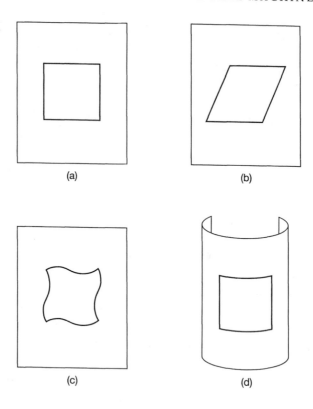

Figure 1.2. (a) A square (2D shape) is drawn in flat 2D space, (b, c) the square can be bent or distorted within flat 2D space or (d) 2D space is itself curved.

means we would have to imagine 4D space in which it could be bent.

If you are still a little confused about the distinction between bending a solid object in space and bending the space itself, here is a simple example in 2D. Take a square drawn on a piece of paper (figure 1.2(a)). The square can be bent *within* the 2D surface (the paper) to form a different shape. For instance, imagine pushing in two opposite corners so that it forms the diamond shape, as in figure 1.2(b), or the lines can be redrawn curved as in figure 1.2(c). This is quite different to the piece of paper itself being bent (figure 1.2(d)). Now the square appears to be bent to us even though we have not redrawn it; rather the space in which the square exists has been bent instead.

2Dworld and 2D'ers

Since it is impossible for us to imagine a higher dimension into which we could curve our 3D world, I will employ a useful trick. We simply make do without one of our spatial dimensions, say the dimension of depth, and then we can deal with an imaginary 2D world (let's be bold and original and call it 2Dworld). Such flat, two-dimensional worlds have been discussed by many authors over the years and have been called everything from Flatland to the Planiverse. The inhabitants of such a universe are flat, cardboard cut-out beings who are restricted to moving not 'on' but 'in' a surface. They can move up/down and left/right but cannot move out of the surface since that would require motion into the third dimension which is impossible for them. Now the illusive fourth dimension that is impossible for us 3D beings to comprehend (but which we would need in order to visualize the curvature of our 3D space) is equivalent to a third dimension as far as the 2D'ers, as I will refer to them, are concerned. We have access to this third dimension even though the inhabitants of 2Dworld cannot.

What would such a 2D universe look like? For a start, the inhabitants would find it just as hard to think about a third dimension as we do trying to think about a fourth. In figure 1.3 are two such beings. It is quite interesting to consider how they carry out basic functions. For instance, their eyes would have to have the freedom to roam about from side to side so that they can see in both directions. If this weren't the case, and the eyes were fixed on either side of their heads then, although they would have the advantage of being able to see in both directions at the same time, they would be missing a vital skill. Being able to look at the same object with both eyes would enable them, as it does us, to judge

9

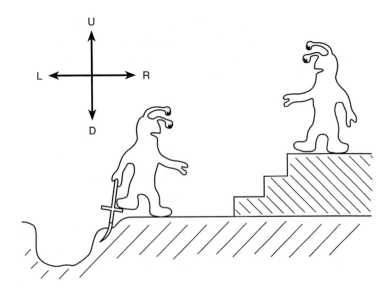

Figure 1.3. Two-dimensional beings living in 2Dworld are free to move up/down and left/right, but do not have access to the third dimension which would involve motion out of the page.

how far away that object is. If they did have both eyes on the same side of their heads, however, they would not be able to look behind them unless they stood upside down! This is because they would be unable to swivel their heads around; a skill that requires access to the third dimension. Both these problems could be overcome if their eyes are free to roam around as I have depicted. Another way, of course, is for them to have a pair of eyes on each side of their heads.

Another problem they would encounter can also be seen from figure 1.3. How does the 2D'er coming down the steps walk past the one digging the hole? He cannot side step him since that would require moving out of the plane (out of their universe) which is not allowed. They would presumably have some kind of convention whereby the one on the left must always give way to the one on the right as in figure 1.4. Or there may be some kind of convention

Figure 1.4. The only way 2D'ers can get past one another. They are unable to side-step each other as that would require one of them to move out of the page.

whereby a 2D'er must always give way to another higher up the social ladder.

A particularly interesting aspect of 2Dworld is what 2D'ers can see when they look at objects in their world. First, let me remind you of what we see when we look at a solid object like a ball. What we actually 'see' is a 2D image on the retina of each eye, which is very important for depth perception. Even with one eye closed we know that what we are looking at is a solid three-dimensional object rather than a flat two-dimensional one, like a disc, due to the way light shining on the ball provides shading. Even without this, we know from experience what a ball looks like and how it behaves. So, when we watch a football match on television we know that the circular object being kicked is a three-dimensional football and not just a flat disc that looks like a ball and is rolling around on its edge. We know this despite not being able to discern any shading on the underside of the ball and despite the television picture itself being a 2D projection of the 3D reality.

When we look at a 3D object we only ever see the two-dimensional surface facing us. Our brains then take into account past experience of such an object plus the way light interacts with that surface to build up a model in our minds of the whole three-dimensional shape even though we cannot see the back of it. How does this compare with what the 2D'ers see? Their equivalent of a sphere is a circle. When a 2D'er looks at a circle she will be looking at it 'edge on' and will therefore only see half of its circumference. She will see on her 'retina' a one-dimensional image: a straight line. Again, she would have to rely on shading to discern the curvature of the line and would have to rotate the circle to be convinced that the line curves all the way round. If the circle is being lit from above, say from a two-dimensional sun overhead, then the top section of the line she sees will be lighter than the bottom section which forms the underside of the circle. Thus, how a circle looks to 2D'ers is not the same as it does to us because they can never see inside it. From our privileged vantage point looking down on 2Dworld we can look inside all objects, not just the circle but the 2D'ers' bodies too. All their internal organs will be visible to us, giving a new meaning to the term 'open-heart surgery'. It is just as impossible for 2D'ers to see inside a closed circle in their world as it would be for us to see inside a hollow ball.

Imagine we came across 2Dworld somewhere within our own universe. In principle, if it were flat then it should extend out forever like an infinitely large sheet slicing through our own three dimensions of space. But let us imagine that it has some finite size and that we came across it somewhere. I don't care where: under your bed, behind your sofa or in your granny's attic. I will assume that we are able to communicate with the inhabitants of 2Dworld[2]. We witness the scene in figure 1.5(a) as a 2D'er attempts to remove an object from inside a square. He cannot even see the object and is not able to get to it without opening the square. For us, not only is the object visible, but we could, if we so wished, reach into 2Dworld and pluck it out of its two dimensions then place it back into 2Dworld outside the square (figure 1.5(b)). We can do this because we have access to the third dimension.

[2] I am assuming that we are able to speak and be heard by them. Sound is transmitted by the vibrations of our 3D molecules of air. Presumably these vibrations would get transferred to the 2D molecules in 2Dworld. All of this is utter nonsense of course, but fun to think about.

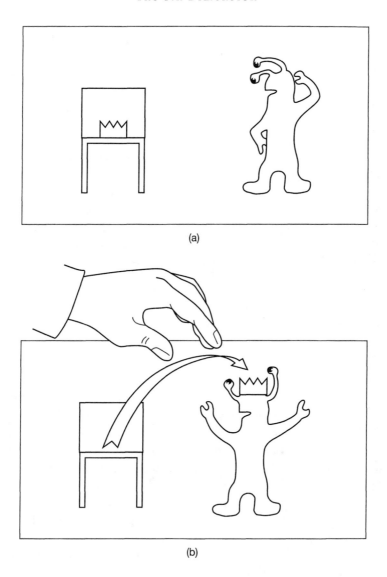

(a)

(b)

Figure 1.5. (a) A 2D'er cannot see a way of getting at the crown locked inside the box without breaking it and setting off an alarm. (b) We can help the thief by plucking the crown out of 2Dworld, into the third dimension, and returning it onto his head.

Having spooked the 2D'er into believing in the power of the paranormal, by causing an object to appear out of nowhere—an object which just a few seconds earlier was locked inside an impenetrable square—we decide to show off the wonders of 3D space by introducing him to a sphere by pushing a small ball into 2Dworld. Of course, it will go right through to the other side provided there is no 2D object in the way. The 2D'er will first see a point growing into a line that gets longer then shorter before disappearing. He concludes from the shading that the line is part of the circumference of a circle and so knows that he is looking at a circle that starts off small, gets bigger, reaches some maximum size (when the ball is half way through) then shrinks again to zero size as it emerges on the other side of 2Dworld. Thus, at any given moment the 2D'er will only ever see a cross-section of the ball.

Curved space

I mentioned that this imaginary 2Dworld need not be infinite in extent and would therefore have an edge, some border defining its boundary. We will see later on that universes do not have edges and so 2Dworld must presumably go on forever. It turns out that this need only be the case (going on forever that is) if 2Dworld is flat, which is what I have assumed so far. What if the inhabitants of 2Dworld lived on the surface of a sphere? Their space is now curved and is no longer infinite in size. After all, a sphere has a certain finite surface area which clearly does not have an edge since the 2D'ers can move anywhere in this universe without ever reaching a point beyond which they cannot go. The important and rather tricky concept to appreciate here is that although 2Dworld is the surface of a 3D sphere, the inside of the sphere and all the space outside the surface need not even exist as far as the 2D'ers are concerned. So, in a sense, the analogy with humans living on the surface of the Earth should not be taken too strongly since we are clearly 3D beings stuck to the surface of a 3D object. The 2D'ers only have access to the 2D surface. The interior of the sphere does not even exist for them.

The interesting question I would like to address next is whether the 2D'ers would *know* that their space is curved.

One way for them to find out would be the way we can prove that the Earth is not flat: by having someone set off in one direction and eventually get back to the starting point coming from the opposite direction having been all the way round the globe. Of course we now regularly send astronauts into orbit who can look back and see that the Earth is round, but the inhabitants of the 2D universe are imprisoned in their surface and cannot move up out of their world to look down on it. There is another way they could check whether their world was curved.

We learn at school that if we add up the values of the three interior angles of any triangle we always get 180 degrees. It does not matter how large or small we draw the triangle or what shape it is; the answer will always be the same. If it is a right-angled triangle then the other two angles must also add up to 90 degrees. If one of the angles is obtuse with a value of, say, 160 degrees, then the other two angles must together make up the remaining 20 degrees, and so on. But before you become too complacent having comfortably negotiated this bit of geometry, allow me to spoil things by stating that this business of angles of a triangle always adding up to 180 degrees is only true *if the triangle is drawn on a flat surface!* A triangle drawn on a sphere has angles which always add up to *more* than 180 degrees. Here is a simple example which demonstrates what I mean. To help you see this you will need a ball and a felt tip pen.

Imagine an explorer beginning a journey at the North Pole. He heads off in a straight line due South (when you are at the North Pole the only direction you *can* head is south) passing through the eastern tip of Canada then down the western Atlantic. He is, of course, careful to steer clear of the Bermuda Triangle since he believes all that superstitious nonsense. He keeps heading south until he reaches the equator somewhere in northern Brazil. Once at the equator, he turns left and heads East across the Atlantic, now moving in a straight line along the equator. He reaches the coast of Africa and carries on to Kenya by which time he has had quite enough of the hot, humid climate and decides to turn left and head

North again. He travels up through Ethiopia, Saudi Arabia, the Middle East, all the way up through Eastern Europe and back to the North Pole.

If you have made a rough trace of his route you will see that he has completed a triangle (figure 1.6(b)). Look closely at the three angles. On reaching the equator and turning left, he had made a right angle (90 degrees). But when he finally left the equator to head back north he made another right angle. These two angles, therefore already add up to 180 degrees. But we have not included the angle he has made at the North Pole with the two straight lines of his outward and inward journeys.. These should also roughly make a 90 degree angle, although of course the size of this angle depends on how far he has travelled along the equator. I have chosen it so that he has traced a triangle, joining three straight lines, with three right-angles adding up to 270 degrees.

Such a triangle is a special case but the basic rule is that any triangle drawn on the surface of a sphere will have angles adding up to more than 180 degrees. For instance, a triangle joining Paris, Rome and Moscow will have angles adding up to slightly over 180 degrees. This tiny departure from 180 degrees is because such a triangle does not cover a significant fraction of the total surface area of the Earth and is thus almost flat.

Getting back to the 2D'ers, they can use the same technique to check whether their space is curved. They would head off in a 2D rocket from their home planet travelling in a straight line until they reach a distant star. There, they will turn through some fixed angle and head off towards another star. Once at the second star they would turn back home. Having traced a triangle they would measure the three angles. If these came to more than 180 degrees[3] they could deduce that they lived in curved space.

Another property, which you may remember from school, is that the circumference of a circle is given by *pi* times its diameter. The value of *pi*, we are told, is not open to negotiation. There is a button on most pocket calculators that gives *pi* up to 10 decimal places (3.1415926536), but most of us remember it as 3.14. OK, I

---

[3] A surface can be curved in a different way such that triangles drawn on it will have angles adding up to less than 180 degrees, but I will come to that later.

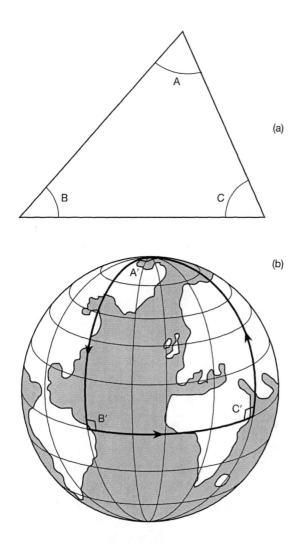

Figure 1.6. (a) A triangle drawn on a flat piece of paper has interior angles $A + B + C = 180°$. (b) A triangle drawn on the surface of a sphere has angles that add up to more than $180°$. Drawn here is one made up of three $90°$ angles.

BLACK HOLES, WORMHOLES & TIME MACHINES

admit that I remember it to the ten decimal places that a calculator shows, but that is only because I use it in my work so often, which is no different to remembering an important phone number. However, I have a mathematician friend who knows *pi* to 30 decimal places. Other than that he is quite normal. We are taught that *pi* is what is called a mathematical constant. It is defined as the ratio of two numbers: the circumference and the diameter of any circle in flat space. If our explorer were to walk round the Arctic Circle, which has a diameter that he could measure with accuracy (it being twice the distance from the Arctic circle to the North Pole), then he would find upon multiplying this value for the diameter by *pi* (which is the way to work out circumferences of circles) he would get a value which was slightly bigger than the true circumference of the Arctic Circle. The Earth's curvature means that the Arctic Circle is smaller than it would be if the Earth were flat.

The properties of triangles and circles that we learn at school are what are known as Euclidean geometry, or 'flat geometry'. The 3D geometry of spheres, cubes and pyramids is also part of Euclidean geometry if they are imbedded in flat 3D space. Their properties change if the 3D space is curved, in a way similar to the way the properties of triangles and circles change when they are drawn on a curved 2D space such as the surface of a sphere. So, our 3D space may well be curved but we do not need to visualize a fourth dimension to 'see' this curvature. We can measure it indirectly by studying the geometry of 3D space and solid objects within it. In practice, we never see any deviation from Euclidean geometry because we live in a part of the Universe where space is so nearly flat we can never detect any curvature. This is analogous to trying to detect the curvature of the Earth by drawing a triangle on a football field. Of course, a football field is not completely smooth. Likewise, space contains regions of curvature here and there as we will see in the next chapter.

What if a fourth dimension of space does exist beyond our three? What can we say about its properties? The best way is to begin by acknowledging that the fourth dimension is to us what the third dimension is to the 2D'ers. Imagine you are standing at the centre of a large circle marked out on flat ground such as the centre circle of a football pitch. If you now walk in a straight line in any direction you will be heading towards the perimeter of the circle. This is called a radial direction because when you reach the perimeter you will have travelled along the circle's radius. On the other hand a bird sitting in the centre of the circle can move along the third dimension: upwards. If it flies straight up then it will be moving away from all parts of the circle at all times.

Now add another dimension to this example and imagine the bird at the centre of a sphere (say a spherical cage). Whichever direction the bird now flies in, it will be moving towards the bars of the cage, and all directions for it are now radial. Just as in the 2D example of the circle where the bird could move along the third dimension away from the circle, we can now see what it would mean to move along the fourth dimension. Starting from the centre of the cage it is the direction in which the bird would have to fly in order to be moving away from all points in the cage at the same time! This is not a direction that we can ever visualize since, as I have mentioned before, our brains are only three-dimensional. So what would we see if we had a magical bird, capable of utilizing the fourth dimension, trapped in a cage? We would see it disappear from view and then rejoin our 3D space somewhere else, possibly outside the cage. It would look as astonishing to us as our 3D skills would look to the 2D'ers were we to pluck objects out of their 2D space.

Another interesting effect of using a higher dimension is what happens when objects are flipped over. Imagine you were able to lift a 2D'er out of his world, turn him over so that his left and right sides are swapped over, then put him back. Things would be quite confusing for him for a while. He will not feel any different but everything around him will be on the wrong side. He would have to adjust to living in a world where the 2D sun no longer rises from the right as it used to, but from the left. And he now has to walk in the opposite direction to get to work from his home.

Things are more amusing if you consider what it would be like for you if a 4D being where to pluck you out of our 3D world and flip you over. For a start, people would notice something slightly different about your appearance since your face now looks to them how it used to appear to you in a mirror. When you next look in a mirror, you will also see the difference. This is because nobody's face is symmetrical. The left side of our faces differs from the right. Maybe one eye is slightly lower than the other or, like me, your nose is slightly bent to one side, or you have a mole on one cheek, and so on. But this is only the start of your problems. Everything around you appears back to front. All writing will be backwards, clock hands will go round anticlockwise and you will now be left-handed if you were right-handed before. One way of testing how things would be like would be for you to go round viewing the world through a mirror. It would take a while before you stopped bumping into things.

Is there really a fourth dimension?

If you already know a little about Einstein's theory of relativity (which I am assuming you do not) then you might well be a little worried at this point. After all, didn't Einstein say something about time being the fourth dimension? In Chapter 6 I will discuss Einstein's theory of special relativity in which time and space are linked together in a quite surprising way, into something called four-dimensional spacetime. For now, we can understand it in the following simple way. Go back to the example of the submarine requiring the three numbers to fix its position. If it is moving, stating those numbers is meaningless unless we also state *when* the submarine was at that position. And so we now require four numbers to correctly locate its position: latitude, longitude, depth and the time when it had those values. However, we should not lose sight of the fact that time is *not* the same as the three dimensions of space. We are free to move forwards and backwards along any one of the three spatial axes, but are restricted to moving forwards only along the time axis (from past to future). The question here is whether there could exist, beyond our senses, a fourth dimension of space.

A hundred years ago, some of the world's most respected scientists believed that the spirit world, the realm of ghosts and phantoms, was four-dimensional and included our 3D space within it. The inhabitants of this higher dimensional world would occasionally pass through our 3D one but would otherwise be invisible to us. Nowadays of course, hardly any serious scientists (by which I mean not counting those of the fruitcake persuasion) believe this. This is not to say that higher dimensions are ruled out. In fact, some new, as yet untested, theories in physics suggested that there may be even more than four dimensions of space, all of which are beyond our grasp. Two theories currently in vogue, known as superstring theory and M-theory[4], suggest that our Universe actually contains nine, and ten, dimensions of space (plus one of time), respectively. But all the extra unwanted dimensions are curled up so small that we can never detect them. You may think this just a load of hogwash but the truth is that either of these exotic theories could well turn out to be the one that describes the ultimate underlying reality of our Universe.

Even if the three dimensions of space we know of are all there is, we will see in the next couple of chapters that it is useful to have an extra dimension up our sleeve to help us understand a certain aspect of Einstein's theories of relativity: curved space.

---

[4] The 'M' stands for membrane, but membrane theory is such a boring name that many physicists prefer it to stand for magic since they claim that the theory is able to explain all the forces of nature.

2
MATTERS OF SOME GRAVITY

Apples and moons

According to myth, Isaac Newton was sitting under an apple tree when, upon being hit on the head by a falling apple, he discovered the law of gravity—implying maybe that the knock on the head produced the flash of insight. (Apple falls—boink—light bulb lights up above head and, hmm, it appears that the ground is exerting a force on the apple pulling it downwards.) Of course it was not that simple. Newton was not the first human to ever notice that things fall down! His insight was much more impressive than that.

It may well be a myth that an apple actually fell on Newton's head, but by Newton's own account it was contemplating a falling apple on his mother's farm (along with other things such as why the Moon goes round the Earth) that led him to his famous universal law of gravitation. What was it about the falling apple that Newton could see that all others before him could not? Stated as simply as possible, he saw beyond the obvious—that all objects had a tendency to want to move downwards towards the Earth— and realized that there was a force of attraction between the apple and the Earth that not only caused the apple to fall downwards towards the Earth but the Earth *to fall upwards towards the apple*. In fact, it is better not to think in terms of objects falling but rather that the Earth and the apple are attracted towards each other.

Outgoing, friendly, popular, family man. All these traits were quite alien to Isaac Newton. Born in Woolsthorpe in Lincolnshire, UK on Christmas day 1642[1], he was a loner who never married and who did not have many friends. He became, later in life, embroiled in lengthy and bitter disputes with other scientists about who had reached certain discoveries first. However, despite the negative image of scientists in general in today's popular media which sadly puts so many teenagers off the subject, Newton was most certainly not a typical scientist. What he lacked in social skills he made up for by being, in many people's view, the greatest scientist who ever lived. He made so many important contributions to so many fields that most of the physics taught at school today is known as Newtonian physics. This is to distinguish it from the modern physics of the twentieth century that will be discussed in this book. Newton also invented the mathematical technique of calculus which is the standard tool for studying most of physics today. The discovery of calculus, however, was the cause of a long-running controversy. The dispute was whether Newton or the German mathematician Gottfried Leibnitz could lay claim to it. In the scientific circles of the time the dispute, in which the English and the Germans each claimed that the other man had stolen their man's ideas, took on a patriotic fervour akin to the modern-day rivalry of the two countries' footballing encounters. However, unlike the modern all too frequent penalty shoot-out resolutions, in the battle of calculus there was no clear winner. Each had, it seems, developed the technique independently. In any case, most of the ground work had been laid down half a century earlier by the great French mathematician Fermat.

Back to gravity. Long before Newton, it was realized that the reason objects fall is because the Earth exerts a force on all things that pulls them towards it. It was also known that the Moon orbits the Earth because the Earth exerts some mysterious force on it stopping it from floating off into space. Newton made

---

[1] This date is according to the Julian calendar in use in Britain at the time. According to the Gregorian calendar which was already in use in other European countries at the time and in use everywhere today, his date of birth was the 4th of January 1643.

the connection between these two phenomena. Attributing the motion of the Moon and the falling apple to one and the same force (gravity) was a bold stroke of genius. Until then it was believed that entirely different laws of nature governed the behaviour of earthly objects (apples) and heavenly bodies (the Moon).

Newton's law of gravity states that any two objects in the Universe will be attracted towards each other by an invisible force. The Earth and each and every object on its surface, the Earth and the Moon, the Sun and the planets, even the Sun and the rest of our galaxy, are all being pulled towards each other. Thus it is not just the Earth that keeps us stuck to its surface; in a sense, we are keeping the Earth stuck to our feet since we are pulling the Earth towards us with as much force as it is exerting on us. When I said earlier that the Earth falls upwards towards the falling apple, I meant it quite literally. It is just that, being stuck to the surface of the Earth, we see the apple moving towards the Earth. But the apple has just as much right in claiming (inasmuch as apples have rights) that it is not moving at all and that it is the Earth that moves towards it.

Likewise, a man and a woman floating close together out in empty space will be *physically* attracted towards each other— even if they are not 'physically attracted' to each other!—by a gravitational force that will cause them to slowly drift even closer together. This force will, however, be very weak (equivalent to the tiny force needed to pick up a single grain of sugar if they had started off a few centimetres apart). The force of gravity is very weak when the masses involved are small.

How is it that the same force of gravity that causes the apple to fall does not pull the Moon down to Earth too? The difference between the two cases is that, despite the Moon's much greater mass, it is in orbit around the Earth and at any moment is moving in a direction that is a tangent to its orbital path, whereas the apple is moving towards the Earth's centre. This is actually a rather bad way of putting it. A better definition of 'in orbit' is to say that the Moon is falling towards the Earth in a curve that forms a circular path around the Earth so that it never manages to get any closer. When Newton first calculated this during the plague year of

1666 he thought he had got the wrong answer and, disappointed, refrained from publishing his results. It was only many years later, when discussing the problem with his friend Edmund Halley (he of comet fame) that he realized the importance of his discovery.

Newton's law of gravity has been tremendously successful for over three hundred years. Note that it is known as a *law* of gravity, since scientists were so sure it was the last word on the subject they elevated it above a mere theory that could be dismissed if and when something better came along. But that is precisely what did happen in 1915. The name was Einstein. Albert Einstein.

Einstein's gravity

Newton's law of gravity would appear to describe an invisible, almost magical, force that acts between all objects however far apart they are (although it does become much weaker with distance) and no matter what lies between them, even empty space. We therefore say that the force of gravity requires no 'medium' (or 'stuff') to act through. Einstein gave a much deeper explanation than this. He claimed that gravity does not act directly on an object but on space itself, causing it to warp. This warping, or curving, of space then causes objects within it to behave in a different way than they would if the space they were in was not warped. Confused? Let us take a step back and see how Einstein came to this seemingly unnecessarily abstruse interpretation.

Have you ever had a ride in one of those amusement park simulators? You take your seat along with a few other passenger inside a closed capsule and watch a short film of a futuristic chase scene. The capsule feels like it is really accelerating, braking, whizzing round sharp corners, riding bumps, climbing and falling. In fact, suspending your disbelief is surprisingly easy. The principle used in these rides is known as Einstein's principle of equivalence and is so simple that it can be stated in one word: *g*-force (or is that two words?). Einstein realized that the force you feel when accelerating (probably felt most clearly when on a plane speeding along the runway just prior to take off) and the force of gravity are equivalent to each other. In fact, we say that the acceleration of the plane which pushes us back against the seat is providing a g-force. The 'g' stands for gravity and is, in fact, a quantity with the units of acceleration not force. So an acceleration of one 'g' would be equal to the acceleration a body undergoes when falling.

At first glance this appears to be rather far-fetched. After all, the force pushing you back in your seat is to do with motion and acceleration whereas the force of gravity acts even when you are standing still (by keeping you stuck to the ground). But think a little about how the simulator ride actually works. How is it that you get the sensation of acceleration even if you look away from the convincing images on the screen? After all, the simulator is not moving anywhere, it just tips and rocks about on its stand. All it needs to do to give the impression of forward acceleration, say at one 'g', is to tip back so that you and your seat are facing upwards. We are so used to the sensation we feel when we lie on our backs in bed at night that we forget about the pull of Earth's gravity forcing our heads down into the pillow. In fact, this force which we usually take for granted is equivalent to the force which pushes us back in our seats if we were in a car accelerating from nought to sixty miles per hour in just over two and a half seconds!

This is why it is so easy to fool the brain into thinking that the gravitational force we are really feeling in the simulator is an acceleration force. In the same way, when our simulator ride stops so suddenly that we feel ourselves being thrown forwards, all that is happening is that the simulator is tipping us forwards and letting gravity do the rest.

Another example that demonstrates the principle of equivalence at work is the flip side of the simulator example, namely using acceleration to simulate gravity. This is the most common example that is used when the subject is taught. Imagine you are strapped in to your seat in a real rocket awaiting countdown for lift-off. Your seat is such that you are facing upwards towards the top (front) of the rocket. Imagine, further, that you are so relaxed and laid back about your trip that you drop off to sleep—not very likely, I know. When you wake up, and before you have a chance to look out of the window, the principle of equivalence would say that you will not be able to distinguish between the sensation you would feel if the rocket were still on the launch pad with gravity forcing you down into your seat, and the sensation you would feel if the rocket had left Earth long ago and was now out in space accelerating at a constant one 'g'. In fact, if you continue to resist the temptation of looking out the window to check whether it is the empty blackness of space or the familiar surroundings of the rocket launch site staring back at you, you would not be able to find any experiment that you could carry out inside the rocket that would allow you to guess where you were[2]. By experiments I mean anything from simple observations, such as studying the swing of a pendulum or watching a ball fall, to sophisticated measurements involving laser beams and mirrors; basically any experiment which could distinguish between the behaviour of objects undergoing an acceleration of one 'g' and the effect of Earth's gravity.

Finally the suspense is too much and you look outside to see that you are indeed accelerating through space. However, all those physics experiments have worn you out so you get back into your seat and go to sleep. When you wake up you feel weightless. You are glad you remembered to strap yourself in or you would have floated off and bumped your head on the instrument panel. Now you are faced with another puzzle if you don't look outside. You see, you could either be drifting in space at a constant velocity with the rocket engines shut off, which would surely account for the sensation of weightlessness, or you could be falling through the Earth's atmosphere and in danger of imminent death if you don't take control of the rocket quickly. You see, when you are falling freely through Earth's gravitational field you experience weightlessness, as though the pull of Earth's gravity has been switched off.

---

[2] OK, in principle, and with sensitive enough equipment, you *could* tell the difference because the Earth's gravitational field is radial rather than planar. This means that if you were to drop two balls side by side on Earth, they would both move along straight lines towards the centre of the Earth. These lines are not quite parallel. In the accelerating rocket they would be exactly parallel.

Free fall

Most of us will never get the chance of being in the above situation in the first place, so here is another example to get this point across.

If you were ever brave (foolish?) enough to bungy jump, you would be forgiven if you felt, as you plummeted towards the surface of the planet accelerating all the time, that the pull of gravity had never been so manifest or more dramatically experienced. In fact, quite the opposite is happening. This may well be the one time in your life that the action of gravity is completely switched off and you are said to be in 'free fall'. For those few exhilarating seconds you are experiencing zero gravity. It is as though gravity has finally got its way and you are doing what it has been trying to make you do all your life. It is just that there is usually solid ground under your feet that ruins things for it. And so, its job accomplished, gravity has temporarily gone AWOL. More correctly, rather than saying that gravity is absent we say that it has been completely cancelled out by your acceleration. The sensation of free fall is what astronauts feel all the time they are floating in space away from Earth's gravity (or in orbit around the Earth) No wonder they have to undergo rigorous training to overcome space sickness. It is a sobering thought to think that space travel is one long bungy jump!

So what does it mean to experience zero gravity? Let's say that, as you fall, you 'drop' a stone that you have been holding in your hand. Since it is falling at the same rate as you it will move alongside you[3]. A physicist's way of viewing this, if she has the presence of mind to stop screaming about how alive she feels and ignores the ground coming up to greet her, is to shut out all her surroundings and imagine that only she and the stone exist. Now the stone appears to be floating in mid-air next to her, in the same way that objects float in zero gravity out in space. This is why, in the rocket example, you would not be able to decide, without

---

[3] This is the experiment that Galileo is supposed to have carried out from the top of the Tower of Pisa—no he didn't bungy jump off—by showing that all objects fall at the same rate no matter what their weight is (as long as they are not so light as to be affected by air resistance like paper or a feather).

looking outside, whether the rocket was moving through Earth's atmosphere in free fall or floating out in space.

Examples such as the ones I have just described are called *thought experiments* since we do not need to physically experience them in order to glean some insights into the workings of nature. Einstein was very fond of such an approach since he spent his time sitting and thinking, rather than working in a laboratory carrying out real experiments. He called these his *gedanken* experiments ('gedanken' is just German for 'thought'). Of course, bungy jumping and fairground simulator rides showing clips from *Star Wars* were not examples he could call upon.

What has all this acceleration stuff to do with Einstein's ideas about curving space? I am afraid I have a bit more explaining to do yet. We must now go back to the example of the rocket. Remember the bit when you wake up and cannot decide, without cheating and looking outside, whether the rocket has yet to take off or is accelerating at one '*g*' out in space? There is a particular *gedanken* experiment you must carry out now. Stand on one side of the rocket and throw a ball horizontally across the rocket, as in figure 2.1(a). The ball will follow a curved trajectory and hit the other side at a point below the one it should have hit if it had travelled in a straight line. This is just what we would expect to happen if the rocket were still standing on the launch pad, with the ball obeying the law of gravity.

If the rocket is now accelerating you should, according to the principle of equivalence, see the ball follow a similarly curved trajectory. Had the rocket been floating freely in space with its engines off (i.e. coasting at a constant speed) it would have carried the ball along with it and you would see the ball move across in a straight line. This is because the ball and the rocket both have the same 'upward' speed. But if the rocket is accelerating, as in figure 2.1(b) (note that the right hand figure is a fraction of a second later than the left hand one), then the ball will not feel this acceleration while it is in flight across the rocket. So by the time it reaches the other side the rocket will be travelling slightly faster than it was when the ball left your hand. The point on the opposite wall where the ball *should* have hit would have moved up slightly and its trajectory

29

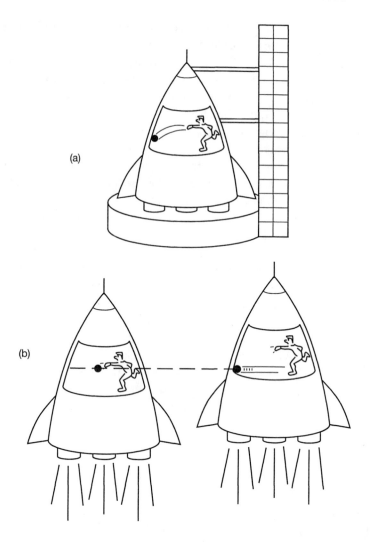

Figure 2.1. (a) The ball thrown under the influence of Earth's gravity will follow a curved trajectory. (b) The ball thrown when the rocket is in zero gravity would have followed a straight line trajectory had the rocket been moving at a constant velocity. But if the rocket is accelerating, as it is here, the thrower will see the path of the ball curve downwards since the rocket will be moving faster than the ball by the time it reaches the other side.

will look curved to you. The principle of equivalence is correct. Although the explanation of the curved trajectory is different in the two cases, the effect you observe is the same.

Next, instead of throwing a ball across the rocket, shine a torch at the other wall so that the light beam is aimed horizontally. If you had sensitive enough equipment you would find that the beam of light bends ever so slightly down towards the back/bottom of the rocket. This is an effect which we can understand quite easily if the rocket is accelerating in space since we would use the same reasoning as in the case of the ball. Although the light from the torch travels across the rocket extremely fast, it still takes a finite time during which the rocket has gained a little extra speed and will have moved forward very slightly.

The problem you might have is believing that the light beam would follow the same curved path when the rocket is standing on the surface of the Earth. But the principle of equivalence is all conquering, and light turns out to be no different to the ball. Even on Earth the light path is slightly curved down by an amount the same as the curvature it has in the accelerating rocket.

Light does not weigh anything[4] so how can it be bent by gravity? However, mass can be thought of as frozen energy, and light certainly has energy, so maybe we can think of it as having weight and should not be surprised if it behaves like material objects and is pulled down by Earth's gravity. In fact, Newton himself had suggested that light is composed of a stream of tiny particles which would be influenced by gravity in the same way as the ball. But I am afraid we would get the wrong answer for the amount of curvature we see if we use Newton's approach. If we were to calculate, based on Newton's argument that light has mass and is pulled down by gravity, the amount of bending we should see in the path of the light beam, we would arrive at an answer that is only half the one we actually measure with our sensitive equipment. Something therefore had to be wrong with Newton's law of gravity, at least when it came to describing the effect of gravity on light.

---

[4] Just accept this for now. I will explain it further in Chapter 6. What I mean of course is that light does not have something called rest mass.

Einstein's reasoning was radically different. His explanation did away completely with the force of gravity. Instead, he said that all material objects in the Universe will affect the space and time in their vicinity causing them to warp. So rather than thinking in terms of the Earth exerting a 'force' on us, apples, the Moon, balls and light beams, which pulls everything towards it, Einstein claimed that the Earth causes the space around it to be curved. Now all objects that move in this region of space are simply following lines of curvature. There is no force that keeps the Moon in orbit and no force that pulls the light beam in the stationary rocket down towards the Earth. Everything moves freely, but along a path that is always the shortest route available. If the space is flat this path would be a straight line, but since the space it moves in is curved so is the path it takes. Such paths in curved space[5] are called *geodesics*.

Einstein developed these ideas during the period leading up to the First World War. He completed this, his general theory of relativity, in 1915. But the world had to wait till 1919 before the theory was verified experimentally.

Einstein had suggested that the Sun's gravity would bend the path of light reaching us from distant stars if the light had to pass close enough to the Sun on its way to Earth. The problem was, however, that when the star is in the same patch of sky as the Sun the bright sunlight makes it impossible for us to see the star. Astronomers had to wait for a total solar eclipse, when the Moon moves between the Sun and the Earth and blocks out the sunlight, to test Einstein's theory. In 1919, the English astrophysicist Sir Arthur Eddington led an expedition to the Amazonian jungle that successfully photographed a solar eclipse and measured the small angle at which the light of a particular star was deflected due to the Sun's gravitational field. It was a difficult and delicate measurement, but it proved that Einstein was right. It made

[5] Yet again, I emphasize that I should be talking about four-dimensional spacetime rather than three-dimensional space alone. Some of the examples and analogies I describe in this book are only to help you get a general feeling for the subject and should not be taken literally. To get a more accurate idea of what is going on is not easy and is beyond the scope of this book.

headlines around the world and Einstein became a household name.

Rubber space

In Chapter 1 I argued that space should not simply be thought of as 'somewhere to put things', but instead that it has its own geometrical properties. These properties are altered in the presence of mass. In order to visualize how space can curve near a massive object we will employ the trick of throwing away one of the dimensions of space and think again about the curvature of a 2Dworld.

The best way to understand what happens to space when we introduce a massive object is to imagine the (2D) space to be like a sheet of rubber. Imagine rolling a small ball across a trampoline. It should go across in a straight line. Now what if you stand still in the middle of the trampoline and get someone to roll the ball again? You will have made a dent causing the trampoline's material to bow down a little. If the ball's path takes it close enough to this dent, it will follow the curvature and be bent round to move in a different direction (figure 2.2). Viewed from above, it would appear as though you had exerted a mysterious force on the ball causing it to be attracted towards you and away from its original straight path. This is how we imagine matter to curve space around it. The curvature causes other objects to follow a different path to the one they would in the absence of the curvature. What has happened on the trampoline is that the ball is following a geodesic path. This is the preferred path for the ball; the one that it wants to take most naturally given the curvature of the trampoline's material that it encounters. Thus, a geodesic path is the shortest distance between any two points. So if you are ever asked what the shortest distance is between two points, don't say a straight line. A geodesic is only a straight line when the space is flat. If the ball had been travelling more slowly along the same path on the trampoline then it would have been caught in the dip and would have spiralled inwards towards your feet.

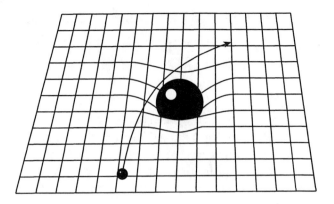

Figure 2.2. Because a massive body like a star or planet makes a dent in space, the paths of smaller objects passing close by will be curved by the 'dip'. This curvature is just what we attribute to the attractive force of gravity.

In the above example, the trampoline's material represents 2D space and so the analogy is only a loose one since all objects in this imaginary space must also reside within the two dimensions, whereas the ball is a 3D object rolling on top of the surface. Similarly the dent that you make by standing on the trampoline is actually due to Earth's gravity pulling you down, whereas I am asking you to imagine that it is *your* mass alone that is curving the 2D surface. In reality, since you are a 3D object in 3D space, what you are really doing is warping the real space around you. This effect is so small, however, that it could never be measured. Nevertheless, it is true that whenever you go on a diet it is not just a flatter stomach that you hope to achieve—something I have found harder in recent years—but the space around you will also be slightly flatter because you have less mass!

So now we can understand Einstein's interpretation of gravity. All material bodies warp space around them, by an amount that depends on how massive they are, and this warped space then guides all bodies that are moving in it, making them travel along geodesic paths. Such paths can be understood if you think about the flight path that an aircraft takes.

A few years ago I flew from London to Tokyo to attend a physics conference. I looked at my world atlas to get a vague idea of the countries I would be flying over. I forgot that a map is a flat projection of the Earth's curved surface. So although the shortest distance between two points on the map (say London and Tokyo) may look like a straight line on paper, to find the true shortest distance we would need to look on a globe. To do this place one end of a rubber band on London and the other on Tokyo. The band will always follow a geodesic line since this will be the shortest distance between the two points. Any other path would be longer and the band would have to stretch more. Since it has a natural tendency to minimize its length it will always find the route which requires least stretching. Now we see that the flight path—assuming the pilot wants to minimize fuel consumption and is not diverted off the geodesic due to bad weather or a country's forbidden airspace—will pass over a region far to the north of both London and Tokyo, a path that looks curved if you plot it on a flat map.

Now that I have introduced Einstein's view of gravity we can go on to look at some of its more fascinating consequences, such as a hole in space into which anything can fall and be lost forever: a black hole. You will discover that such fantastic objects are science fact not fiction because astronomers are now almost certain that black holes really exist out in space.

To pave the way for a discussion of black holes we must first learn a little about how they can form. For this to happen, space needs to be warped by an incredible amount. This requires something very dense indeed. Even the whole Earth is not enough—which, by the way, rules out any possibility of the Bermuda Triangle being some kind of hole in space that swallows up unsuspecting ships and aircraft, since a hole of that size would require a mass much more than that of the whole planet, and we can easily work out the mass of the Earth from the way it orbits the Sun.

What we need for some serious warping of space is something big, such as a star.

Twinkle, twinkle

What emotions are conjured up when you look up at the sky on a cloudless night? Are you awe-struck by the vastness of the heavens? Have you wondered what is happening up there, among those sparkling pinpricks of light we call stars, so impressive in number yet each appearing so insignificant? It is easy to forget what they really are: gigantic cauldrons of fire, millions of times bigger than the Earth. For stars are so distant that it would take our fastest rockets many years to reach even our closest neighbour. But there is a star that our rockets could reach in a matter of months.

The closest star to Earth is an old friend. Without it we would not be here. Our Sun sustains nearly all life on Earth with its warmth and light. The heat generated inside it and radiated out to bathe its planets is something we all take for granted. The author Douglas Adams sums up our indifference wonderfully:

> "*Several billion trillion tons of superhot exploding hydrogen nuclei rose slowly above the horizon and managed to look small, cold and slightly damp.*"

Stars come in many different colours and sizes and our Sun is pretty average. It is middle-aged and rather on the small side. An astrophysicist will tell you that it is a yellow, main sequence, G2 dwarf star. Doesn't sound very impressive does it? Almost a bit embarrassing. You can just imagine the intergalactic snobbery as visiting aliens from big, white class A stars like Vega or Sirius look down their ears at us (their noses being on top of their heads). But in the domain of stars, being small has its advantages.

All stars have a certain life expectancy which can vary from a million to many billions of years. It all depends on what is going on inside them and this depends ultimately on their mass, which is a measure of how much matter they contain. So what goes on inside a star? We now know that all stars are like cosmic cooking pots. Most of the atoms that make up your body were synthesized inside some star long before the Sun and our solar system were even made; a star, moreover, that no longer exists. I am aware that it appears we are being side-tracked from our main story of how a black hole is created, but the life cycle of a star is crucial to this story. Stars go through several quite different phases during their lifetime, each more fascinating than the last.

Cooking the elements

Everything around us is made up of atoms. These come naturally in ninety-two different varieties, called the elements. They range from the very lightest gases, such as hydrogen and helium, then carbon, oxygen, nitrogen and heavier elements such as aluminium, nickel, iron, gold and onto the big boys like lead and uranium. Have you ever wondered how these different atoms came to be made in the first place? The process is known as nucleosynthesis—try saying that three times quickly. Less than a minute after the birth of the Universe conditions were such that the lightest two elements could be synthesized and the Universe thereafter contained roughly 75% hydrogen and 25% helium, with a seasoning of the next few elements in the periodic table such as lithium and beryllium. This concoction is the raw material of stars. When clouds of this interstellar gas form, they begin to contract under the influence of their own gravitational attraction. As the gas becomes denser, it heats up and, slowly, a new-born star is formed at the centre. When this temperature reaches a scorching few million degrees, conditions become hot enough for the star to switch on.

Stars shine due to the process of thermonuclear fusion. This is when the nuclei of two hydrogen atoms fuse together to form the nucleus of a helium atom, releasing in the process a vast amount of energy. Scientists have been trying, unsuccessfully so far, to mimic this process on Earth in a controlled way to produce an unlimited, clean (in the sense of not being radioactive) energy supply. The problem is, of course, that we cannot stop the extremely high temperature plasmas in our fusion ovens from escaping. Stars, on the other hand, continue to burn and shine brightly all the time the fusion reactions are going on inside them because their gravity keeps them together. At the same time, this process provides an outward pressure that keeps at bay the crushing inward pressure of the star's gravity.

This has been going on inside the Sun for the past five billion years since it was born (along with its nine planets) from a cloud of gas and dust. The Sun will continue to shine happily like this for a further five billion years. So it is roughly half way through its life at the moment. As far as stars go, this is an impressively long lifespan, for which it has its small mass to thank. The more massive a star is, the stronger its gravitational pressure will be, and so the denser and hotter its interior becomes, and the faster it burns its nuclear fuel. The very largest stars, a million times the mass of the Sun, will live for just a few million years.

Five billion years from now the Sun will begin to run out of its hydrogen fuel and will gradually move into a new phase of its life. It will become something called a red giant star. When it uses up all the hydrogen in its core it will begin to collapse under its own weight and all the matter in the core will become compressed and so heat up again. At this point, two very different things happen. First the heat in the core is such that helium atoms are now forced together to make heavier elements. At the same time the outer layers of the Sun expand and swell up to such a size that the closest planet to it, Mercury, will be swallowed up. The Sun will now be many times brighter than it was before, and will fill up half the sky as seen from Earth. Unfortunately, we will not be able to witness this event since the surface of the Sun would now be so close it would vaporize the Earth. In any case, if humans are still around five billion years from now they will, hopefully, have long since found a new home.

After a further billion years the Sun will enter the final phase of its life by shedding some fraction of its contents out into space. This forms a rather pretty disc of gas called a planetary nebula, at the heart of which will sit the Sun's dying core: a white dwarf star. Such an object forms when the bulk of the Sun's mass has collapsed in on itself due to its own gravity when the processes of thermonuclear fusion finally cease. It will comprise mainly of crystallized carbon and oxygen and will resemble a massive spherical diamond the size of the Earth. Gradually, this white dwarf will cool and become dimmer and colder until it finally goes out completely. Such an object is extremely dense and just a pea-sized fragment of it would weigh about a ton.

Thus will our Sun end its days rather unremarkably, even ignominiously, when compared with many bigger stars which can lay on an impressive fireworks display.

Champagne supernovae in the sky

Not all stars end their lives as white dwarfs. In fact, if a star is more than a few times the mass of the Sun it is destined for a much more spectacular end. Once the nuclear processes inside it cease, its extra mass means that it will exert more gravitational pressure on its core. This causes the core to become so dense and hot it sends a shock wave of matter back out through the star causing it to explode as a supernova. Briefly it will be the most spectacular object in the whole galaxy. For a few days it will shine a hundred million times more brightly; brighter than all the other stars in the galaxy put together.

One property of stars I have not mentioned is that most of them come in pairs, called binary systems, in which the two stars orbit around each other. In fact, single, isolated stars such as the Sun are in the minority.

The above scenario of a single massive star exploding is known as a type II supernova. These have varying degrees of brightness and do not depend on whether the star was part of a binary or not. There is a more common way a star can go supernova. It is known as type I, and occurs in binary systems. Even if a star is not initially massive enough and ends up as a white dwarf, it may still be able to suck material from its companion star and put on weight. It can thus gain the critical mass this way.

One of the most celebrated supernovae in recent years was seen in 1987. All the stars we see in the night sky are in our own Milky Way Galaxy. Other galaxies are so far a way that we cannot see individual stars. The star that exploded in 1987 was not in our galaxy but in a neighbouring one known as the Large Magellanic Cloud. Yet, at its brightest, it could be seen clearly in the night sky.

At the centre of many supernovae remnants resides a small dense core, the remains of the original star. This object has the same diameter as a large city such as London or New York, so it is much smaller than a white dwarf star. It is therefore far denser since it still contains a significant fraction of the material of the original star that exploded. A tiny piece of this dense core the size of a pea would weigh, on Earth, as much as Mount Everest! Such an object is called a neutron star, and is one of the most fascinating objects in astrophysics. In fact, neutron stars are the subject of much current research activity. You may also have come across the term 'pulsar'. All neutron stars spin very rapidly and sweep a beam of radiation out into space as they do so. If the Earth happens to be in the path of this sweeping beam, the neutron star will look to us as though the light is pulsing on and off, hence the name pulsar. Some pulsars spin many times per second and I will be coming back to them later in the book when I consider the possibility of using them to make a time machine.

Despite all these exotic sounding astronomical objects, we have yet to meet a black hole. Let us consider what happens when an even bigger star, say twenty or thirty times the mass of the Sun, stops shining. Such a star will not be able to resist its own gravitational collapse. It will keep on collapsing until it has been squashed to such a density that even its own light cannot escape its gravitational pull. To someone watching from a distance the star will suddenly disappear from view. It has become a black hole.

But there is much more to the story than that and I will come back to black holes in Chapter 4. In the next chapter we will put to use some of the ideas about the curvature and stretching of space to look at the Universe as a whole. A lot of what we have learned about the Universe has only become known over many years of astronomical measurements and observations. Some theoretical ideas have yet to be confirmed while others remain highly speculative. One thing is for sure: there are still many unanswered questions. Over the next few pages I will review some of the latest ideas about the origin, shape, size and fate of our Universe.

3

THE UNIVERSE

The Universe may be closed, but it opens again after lunch.
Erica Thurston, *Surrey physics student*

The night sky

If, like me, you live in a densely populated urban area, where
light pollution means that even on a clear night you can only see
a handful of the very brightest objects in the sky, then you will
probably be hard pushed to recognize many of the stars or planets.
I can still pick out Mars and Venus, our closest neighbours apart
from the moon, but I am no longer sure about the constellations.
As a child, I use to be much more familiar with the night sky.

I was born in Baghdad and spent the first sixteen years of
my life in Iraq, but left the country for good with my family in
the late 1970s when the political climate changed. Prior to that, we
would visit England every two or three years to spend our summer
holidays with my grandparents. However, the Iraqi summers held
their own magic. The last of the clouds would have dissolved
away by late April, destined not to blemish the blue skies again till
October, and school holidays would stretch out for three and a half
glorious hot months (we did have a six day school week though).
During July and August, temperatures would reach a maximum
in the mid-forties (degrees Celsius) and would hardly ever drop
below an uncomfortably sticky thirty degrees at night.

The most exciting ritual confirming the arrival of the Middle
Eastern summer was when the bedding was carried up to the roof.

41

Houses had stairs leading up to flat roofs where everyone slept for roughly a quarter of the year to escape the stifling heat and humidity. Summer nights thus always meant lying awake gazing up at a sky teeming with thousands of stars, trying to make out shapes and patterns by joining the 'dots'. Eventually they would be partially obscured by the mosquito nets that we would pull over the beds enclosing each of us in our own transparent tent. There was never the remotest concern that it might rain. It never did in the summer.

Living now in Southern England I have almost forgotten how beautiful the night sky can be, and I sometimes miss the thrill of watching out for shooting stars.

So, yes, I use to be able to recognize a few of the stars. Lying up on our roof as a young child I learned that some of the brightest 'stars' were not stars at all but planets, which only shone because, like the Moon, they reflected the light of the Sun when it was on the other side of the Earth. The true stars were millions of times further away than the planets, and so had to be shining many times more brightly for us to be able to see them. I also vaguely remember being both a little disappointed and exhilarated when I found out that a shooting star was nothing more than a tiny rock burning up as it entered Earth's atmosphere, and that it was really called a meteor.

This chapter is a mix of two related scientific fields: astronomy and cosmology. Most people have a good idea what astronomy is about, but not many are familiar with what cosmology means. As far as 'ologies go, you must agree that cosmology sounds pretty impressive and awe-inspiring. It is the study of the whole Universe: its size and shape, its birth and evolution, even its likely fate. It is also seen to be the most glamorous area of physics. It addresses, and even professes to answer, questions which many feel are beyond the realm of science.

Most of what we now know about the Universe has come about through painstakingly careful experiments and astronomical observations, which are continually being refined as more powerful telescopes are built and new techniques developed. But while cosmology is, loosely, a sub-field of astronomy, the knowledge we have gained about the Universe has also come from other areas of science, such as nuclear and particle physics and theoretical astrophysics. Theoretical cosmology involves creating idealized mathematical models of the Universe by solving the equations of Einstein's general theory of relativity. These can be formulated in such a way so as to describe the properties of the whole Universe and not just a small region of space and time in the vicinity of a massive object, such as a star.

As in other parts of this book, I will be discussing ideas about our Universe which, at the time of writing at least, represent our best current understanding and favoured theories. A few years from now some of these may well be shown to have been wrong. On the other hand, there are certain properties of the Universe that we are pretty confident about and which I am sure will stand the test of time. At the end of this chapter I will summarize which features of the Universe are, in my view, correct and which are still open to debate.

To give you an idea of just how rapidly ideas and theories in cosmology are changing and advancing due to ever more accurate astronomical measurements, I had to rewrite substantial chunks of this chapter during the proof-reading stages of the manuscript. In fact, we shall see that 1998 was an important year in cosmology.

How big is the Universe?

I am tempted to just say VERY BIG! and leave it at that. In fact, according to the most recent astronomical evidence, it may well turn out that the Universe is infinite. This means that it just goes on forever. However, we can only ever see a small part of it, even with the most powerful telescopes we could hope to build. There exists a sort of horizon out in space beyond which we can never see that marks the limit of what is known as the *Visible Universe*. This is not a real edge but has to do with the fact that the Universe hasn't been around forever and light takes a certain time to reach us. I will go into this a little more when I discuss something called Olbers' paradox.

The Earth orbits the Sun at a distance of 150 million kilometres, which is equivalent to almost 4000 times round the Equator.

Together, the Sun and its planets form the solar system. Earth's orbit takes 365 days and six hours, which is why we need leap years of 366 days, since four lots of six hours will make up the extra day.

Of course it makes no sense to measure vast astronomical distances in kilometres. Instead they are measured in terms of the distance light travels in one year. In the chapter on special relativity we will see that the speed of light is the fastest speed attainable by anything in the Universe[1]. But it nevertheless takes light a certain time to get from A to B; it just depends how far away B is. This may not be obvious to us when we flick a light switch in a room. To us, the whole room is instantly bathed in light, but this is only because the distance light needs to cover from the bulb to the four corners of the room is so small. In fact, it takes the light typically only ten *billionths* of a second to get from the bulb to the walls of a room.

Over astronomical distances, the time taken by light to travel from one place to another becomes appreciable. For instance, it takes light from the Sun eight minutes to reach the Earth: just eight minutes to cover 150 million kilometres. But it takes five hours to reach the outermost planet, Pluto. In a whole year light could cover the distance from the Sun to the Earth sixty thousand times. This distance that light can travel in one year is known, imaginatively, as a *lightyear*. (Well, what else would you call it?) It is nevertheless a little confusing to use a term containing a time span to define a distance, but there you go.

These vast cosmic distances mean that cosmology has a clever trick up its sleeve. When we look through our telescopes at a star that is one lightyear away, we must remember that what we are seeing is the light that left the star one year earlier. So we are not seeing the star as it is now but a slightly younger version of itself. In effect, we are looking into the past. In geology and archaeology, scientists look at the evidence around them (rocks and fossil remains) and try to infer what things were like in the

---

[1] Hypothetical particles known as tachyons, which would travel faster than light, are predicted by Einstein's theory of relativity but probably don't exist in the real Universe.

past. Astronomers, however, can look directly into the past. The further out in space they look, the older the light their telescopes are picking up and the further back in time they are probing. The very furthest objects that can be detected from Earth are billions of lightyears away and show what the Universe was like when it was very young.

Apart from the Sun, the closest star to us is a much fainter and smaller dwarf star called *Proxima Centauri* which is just over four lightyears away. Relatively close to this star is the *Alpha Centauri* binary system, which is a pair of stars similar to the Sun that orbit each other once every eighty years. Incidentally, *Beta Centauri* is nowhere near *Alpha Centauri* but a hundred times further away. It is just that, being a very bright giant star, it shines with a similar brightness in the same region of the night sky and so they look, to us, to be close together.

Stars are so far apart that you would be right in thinking that most of space is just that: space. But you would be wrong in thinking that the stars are spread out *evenly* throughout the Universe. The distances to our closest neighbours that I have quoted above are quite typical between stars in our neck of the woods, but elsewhere stars can bunch up much more closely, and there are vast stretches of the Universe which contain no stars at all. Without exception, all stars congregate in large groups called galaxies. We live in the Milky Way Galaxy (with a capital G to distinguish it from other galaxies), which is shaped like a flat disc with a bulged-out central region. The visible outer region consists of spiral arms which give it its name: a spiral galaxy. It is eighty thousand lightyears across and, to give you some idea of this size, there are more stars in our Galaxy (about a hundred billion) than there are people living on Earth (about six billion). The Sun is situated towards the edge of the Galaxy, on one of its spiral arms, and orbits the centre of the Galaxy once every 255 million years. The galactic centre is much more densely populated and contains stars which are older than the Sun.

It is helpful to think of the Galaxy as a great star city, with the Sun situated out in the modern suburbs, far from the hustle and bustle of the downtown galactic centre. All the stars we see in the sky with the unaided eye are in our Galaxy, but there are many *billions* of other galaxies, each with its own population of stars. Very few of these stars, even in neighbouring galaxies, can be singled out even with a telescope. The only time one can be seen with the naked eye is if it undergoes a supernova explosion when it briefly outshines the rest of the stars in its galaxy put together.

Not only do stars cluster together in galaxies, galaxies also group together in clusters. Our Galaxy is one of a motley collection of about forty, known as the *Local Group*. Closest to us are a number of dwarf galaxies hanging on to the coat tails of our own. The nearest large galaxy to us is the *Andromeda Nebula*, which is about two million light years away and is the only galaxy, beyond our own, that is clearly visible from Earth with the naked eye.

Astronomical measurements have reached such a degree of precision and sophistication with ever more powerful telescopes being built, allowing us to probe ever deeper into space, that we now know that galaxy clusters are themselves grouped together into what are known as superclusters. Our Local Group is in fact part of the *Local Supercluster*. What next? A cluster of superclusters?

What does all this tell us about the Universe? For one thing, it is very lumpy. On every scale: from stars to galaxies to clusters to superclusters, matter tends to clump together unevenly. This is, of course, due to the all conquering force of gravity which dictates the structure of the whole Universe. The mutual gravitational attraction of all the stars in the Galaxy keep them bound together. It is gravity that causes galaxies to clump into clusters and superclusters, and the gravitational pull of all the matter in the Universe that dictates its overall shape.

The expanding Universe

These days, many non-scientists will have come across the concept of the expanding Universe. But what does it mean? Is it just another strange idea devised by scientists based on a scrap or two of evidence that could just as well have been interpreted in another way? The answer is no. There is now so much evidence in support of the observation that our Universe is getting bigger that we can no longer be in any serious doubt. The expansion was confirmed as long ago as 1929 when the American astronomer Edwin Hubble made a remarkable discovery, but only after several cosmologists had already predicted the effect theoretically.

The first modern day cosmologist was, of course, Einstein himself. Soon after he completed his general theory of relativity in 1915 he began using his equations to describe the global properties of the whole Universe. He soon came across a serious problem. If, at a given time, all the galaxies in the Universe are stationary relative to each other, and provided the Universe was finite in size, then their mutual gravitational attraction will cause them to begin to converge on each other and the Universe would collapse in on itself. It could not remain static. This is actually quite a tricky concept to come to grips with (and not the only one we will encounter in this chapter). This is because, naïvely, you would think that the Universe as defined by its volume of space should remain the same size while the matter it contains gravitates towards its 'centre'. This is quite wrong. For one thing we will see that the Universe does not have a centre at all and, in any case, we have learnt that gravity affects space itself and does not simply act on the matter 'within' it.

The prediction of his own equations bothered Einstein. The widely held view at the time, and Einstein was no exception to this despite his many other revolutionary ideas, was that the Universe, at the level of galaxies and larger, should be static and unchanging. Whether it had been thus for ever or whether a divine creator had conjured it into existence at some time in the distant past did not matter. Both views supported a picture of the present Universe that was constant. The idea of an evolving universe was both alien and unnecessary. So, when Einstein's equations of general relativity seemed to indicate that the Universe should be shrinking he decided to patch things up. He argued that, in order to balance the inward pull of gravity there needed to be an opposing force of antigravity, known as the cosmic repulsion force, which would balance the gravitational attraction and keep all the galaxies apart and the Universe stable. The difference between gravity and antigravity is the same as the difference between the attractive force that pulls the north pole of one magnet towards the south pole of another and the repulsive force that pushes two north poles apart. This cosmic repulsion force appeared in the mathematics as a number, which Einstein called the *cosmological constant*. It was denoted in his equations by the Greek letter *lambda*. (In advanced mathematics it is not enough to use x, y and z for the unknown quantities. We soon run out of letters in the alphabet and start raiding Greek letters—with *pi* being the best-known example of this). What Einstein had suggested was a mathematical trick in order to achieve his model of a static universe.

A few years after Einstein's initial work, the Soviet cosmologist Aleksandr Friedmann published a paper in which he suggested doing away with the cosmic repulsion (by setting the value of the cosmological constant to be equal to zero in Einstein's equations). Friedmann found that when he applied Einstein's equations of general relativity to the Universe and worked through the maths, he always found solutions (other equations) which predicted that the distance between any two points in space was stretching over time. He had found theoretically that the Universe was getting bigger all the time. Two other scientists came to the same conclusions round about the same time. They were the Dutch astronomer Willem de Sitter and the Belgian cosmologist (and priest) Georges Lemaître.

This may seem rather surprising if we think what the action of gravity would be when there is no cosmic repulsion force to hold the matter in the Universe apart. Surely, without a cosmological force of repulsion the Universe should be shrinking not growing. But an expanding universe can be understood in the following way. Imagine that something had set the Universe expanding in the first place, an initial explosion. The gravitational pull of all the matter in the Universe would then be trying to slow the expansion rate down. This was the essence of Friedmann's argument. If there is no cosmic force of repulsion to balance the attraction of gravity, and the Universe had started off expanding, then it would have to be either expanding or contracting at the moment. It could not remain poised between expansion and collapse since that would be unstable.

A simple example to demonstrate this would be what happens to a ball on the side of a smooth slope. If placed half way up the slope it will always roll down. But if we did not see how the ball came to be on the slope in the first place then we would expect it to be either rolling up the slope (corresponding to an expanding universe) or down the slope (a collapsing universe), never standing still. Of course the only way it could be rolling up the slope would be if it had been deliberately given an initial push, but in that case it would immediately begin to slow down and eventually start rolling down again. Now imagine that the slope levels off at the top. Provided the ball was initially set rolling up the slope fast enough it could make it to the top. Once there, it could then continue to roll along indefinitely without slowing down (of course I am ignoring friction and wind resistance here since a real ball would eventually stop even on a flat surface).

Assuming that the ball was always given the same initial speed up the slope, what governs its ultimate fate will then be how high the top is. If it is too high, the ball will not be able to reach the top and will roll down again.

This is how we can view the expansion of the Universe. The effectiveness of the gravitational pull depends on the amount of matter the Universe contains. By matter I do not just mean all the stars, planets and other solid objects, but *everything* of substance in the Universe. This may be in the form of dust, gas, subatomic particles, even pure energy. So, whether the Universe is now contracting or expanding depends on how much matter it contained and how long the gravitational pull of all this matter had been applying the brakes on its initial expansion. This was the essence of Friedmann's model universe.

No one, not even Einstein, was prepared to believe Friedmann's results, not until experimental proof was found. This came just a few years later. Unfortunately, Friedmann died in 1925 and did not live to see it.

Hubble, bubble...

Edwin Hubble almost became a professional heavyweight boxer. Instead he chose a career path in astronomy, and now he has the world's most famous telescope named after him. What was his claim to fame? For one thing, he was the first to realize that other galaxies existed beyond the Milky Way. Until then, it was thought that the tiny smudges of light that could be seen through telescopes were clouds of dust, called nebulae, within our own galaxy. Hubble found that they were too far away to be part of the Milky Way and therefore had to be galaxies in their own right. He also discovered that these other galaxies appeared to be flying away from our own at speeds which were proportional to how far away they were. The further away a galaxy was, the faster it appeared to be receding from us. What was remarkable was that this was happening whichever direction he pointed his telescope. He had shown experimentally that Friedmann's model of the expanding Universe was correct. Einstein was forced to admit that including the cosmological constant in his equation had been the biggest mistake of his scientific career.

Hubble argued, correctly, that since the Universe is now expanding, then in the past it must have been smaller than it is today. Imagine that the expansion of the Universe could have been filmed from a vantage point that has to be somehow 'outside' the Universe—of course this is impossible since all of space is, by definition, within the Universe. By running the film backwards you would see the Universe shrinking. If you went back far enough into the past you would reach a time when all the galaxies overlapped each other and things would have been pretty crowded. Go back even further in time and all the matter will get more and more squashed up and squeezed closer together until you reach the moment of the Universe's birth, the Big Bang[2].

Hubble made his discovery by measuring something called the *cosmological redshift* of light. To understand what this means consider a more familiar phenomenon called the Doppler shift

[2] The term 'big bang' was not coined until the 1950s by the astrophysicist and author Fred Hoyle.

which, as you probably know, is the change in pitch you hear when, say, a fast ambulance goes past you. The reason for this effect is the change in frequency of the sound waves which reach you from the ambulance when it is in two different situations: moving towards you and moving away from you. When it approaches, the waves of sound get squashed up, giving rise to a higher frequency (high pitch) but when it is receding the waves are stretched out to give a lower frequency (low pitch).

The same thing happens to light. When an object is moving away from us—say a distant galaxy—the waves of light that reach us from it get stretched and the frequency of the light goes down. Instead of the frequency of the light we more often talk about its wavelength. You probably remember something about wavelengths from your school physics. You know, ripple tanks, long springs that stretched across the class. What fun! Anyway, the wavelength is the distance between two consecutive wave crests. So a drop in frequency of light is really due to the stretching of the wavelengths.

Since we are confident that a distant galaxy should contain stars similar to the ones in our own Galaxy, and since we know what wavelength the light should have—the nuclear processes inside all stars cause them to shine with light of specific wavelengths—then by measuring the change in wavelength of the light we can work out how fast the galaxy is moving away from us. Of course, astronomers will be quick to point out that it is not as simple as this, but the basic principle is correct. I will come back to some of the subtleties of measuring the rate of expansion later on.

The effect is called a redshift because the wavelength gets stretched as the galaxy recedes, and longer wavelengths of visible light are associated with a redder colour. The term 'redshift', while only really applying to visible light, is nevertheless used for all parts of the electromagnetic spectrum.

We must first consider whether this reddening of light from the distant galaxies observed by Hubble could be interpreted in another way. Astronomers certainly tried to since they did not initially want to believe that the Universe was really expanding.

An obvious way this could happen is if light loses energy on its way from its source to our telescopes, since a decrease in energy would also make the wavelength longer. The only way the light would lose energy would be if it was having to fight its way through any interstellar dust or gas it encountered while on its long journey through space. But there was a fatal difficulty with this explanation. Light loses energy by bouncing off the atoms of matter in its path. So it would tend to move in a zigzagged path and this would make the image of the galaxy appear blurred. Since there was no observed blurring of the images of the galaxies this explanation had to be ruled out. The only other explanation was the Doppler shift due to an expanding universe. A few physicists, including a colleague of mine who taught me relativity as an undergraduate, argue that the redshift can be explained by something known as a transverse Doppler shift. This is the Doppler shift observed in the light from objects moving at high speed across our field of vision and not away from us. This is quite correct. However, I will show that the redshift is not the only evidence we have of expansion.

Space is stretching

Let us take a closer look at what Hubble's discovery means. How can all the galaxies be receding away from *ours*? Surely this means we must occupy a privileged position in the Universe. We must be sitting exactly at its centre. If galaxies on all sides of us, and which are the same distances away from us, are travelling away at the same speed then we might conclude that we are not moving at all. It is as though all the matter in the Universe originated from our little corner of it.

It may be that we are unique in being the only life in the Universe, although even this seems pretty unlikely given the sheer size of the Universe. But we most certainly have no reason for believing that we occupy a privileged location in the Universe. In fact, an important tenet in cosmology, known as the cosmological principle, says that there is no preferred place in the Universe.

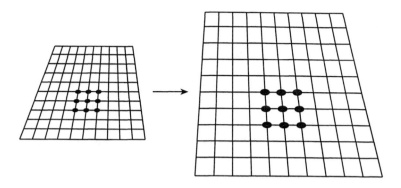

Figure 3.1. The rubber sheet model of expanding space. Imagine the 2D galaxies evenly separated on a grid. As the sheet expands, each galaxy will see all the surrounding ones moving away from it.

That, on the very largest scale, the Universe looks the same everywhere. So how does everything appear to be moving away from *us*?

The answer is deceptively simple. It is not that the galaxies are flying through space away from our own, but rather that *the space in between is stretching*. Imagine a large sheet of rubber on which you place markers on square grid points so that they are all at equal distances from each other (figure 3.1). If the sheet is stretched equally in all directions the distance between any two markers will increase. Every marker would see all surrounding ones moving away from it and no one marker is any more special than the others. Of course I am assuming the sheet is very large, otherwise we would have to worry about those markers that are sitting out on the edge.

When I lecture on this topic I almost always get some bright spark in the audience who asks the following question: if space is expanding and everything is imbedded in space then surely everything should be expanding together, including us and all our measuring equipment on Earth. If the distance between our Galaxy and another one doubles over a certain period then, surely, the distance between all the atoms in our bodies, measuring tapes and rulers will also double. So how would we ever 'observe' the expansion?

Once they have asked this question they tend to turn triumphantly to the rest of the audience as if to say 'there, let's see him get out of that one!'

However, the answer is surprisingly simple. Remember that gravity acts to slow down the expansion of space, and if gravity were strong enough it would win over the expansion completely. At the level of the whole Universe the expansion rate is high and the density of matter very low. But at the level of our Galaxy, the space within it will not be affected since gravity is strong enough on this scale not to permit any expansion. Down at the level of humans and our measuring devices, matter is densely packed together and the atoms that everything is made of are held together by a force much stronger even than gravity. It is called the electromagnetic force and is the glue that binds atoms together. Space is most certainly not allowed to expand at this level and so we, and everything else on Earth, remain the same size.

Time for an everyday example of this (please skip this paragraph if you are already convinced by the previous one). Consider the air bubbles that rise up from the bottom of fish tanks. These bubbles start off small because the pressure of the water at the bottom of the tank is high and squeezes the air inside the bubble. As the bubble rises, the pressure decreases and the bubble expands due to the outward push of the air molecules inside it. Since the number of air molecules inside each bubble does not change, they must be further apart when the bubble is large. However, and this is the crucial point, we would not expect each molecule of air to grow in size along with the bubble.

An interesting point to make is that the nearest galaxy to our own, Andromeda (or M31 to give it its highly imaginative astronomical name), is actually moving towards us! Andromeda is two million lightyears away and, according to current estimates of the expansion rate of the Universe, should be moving away from us at a speed of fifty kilometres per second. Instead it is moving towards us at three hundred kilometres per second! The expansion of the Universe, therefore, does not even show up at the scale of our Local Group of galaxies, let alone on Earth.

Hang on a minute, you might be thinking, did Hubble get it wrong after all? I thought he observed *all* galaxies moving away from ours? The answer is that galaxies are not uniformly distributed with equal spacing throughout the Universe. Hubble had been observing very distant galaxies, that *are* moving away from us, and not the nearby ones.

The speed at which our Galaxy and Andromeda are moving together is equivalent to covering a distance all the way round the world in two minutes, or a distance between the Earth and the Sun in under a week. But before Hollywood decides to base its next blockbuster movie on how a few brave men and women save the Earth from an imminent collision with Andromeda I should point out that, at the current rate, it will take several billion years for the two galaxies to merge. Even when this eventually happens it is highly unlikely that anything will actually hit the Earth since, as we have seen, stars are quite far apart and the chance of a star ploughing through the solar system is remote. Physicists are able to build sophisticated computer simulations that show dynamically how two galaxies behave when they merge together.

So what of Einstein's antigravity force, his cosmological constant that appeared in his equations to stop the Universe from collapsing under its own weight? Did the discoveries of Friedmann and Hubble consign it permanently to the scientific scrap heap?

As the field of cosmology has evolved and matured over the twentieth century, the cosmological constant has proved to be rather more durable and resilient. In fact, it has made more comebacks than the Rolling Stones[3]. For a while, cosmologists decided that it would not and should not be entirely ruled out in Einstein's equations. Maybe it should be left in but be given a very small value so that it did not conflict with Hubble's observations. Remember I am talking here about an abstract mathematical model of the Universe that is predicted when Einstein's equations are solved. By varying the value of the cosmological constant, cosmologists could then study the properties of the different model

---

[3] Of course if you are old enough to be a Stones fan you would argue that they have never been away and so have never needed a comeback.

universes predicted. These properties could then be compared with those observed in the real Universe.

Upper limits were computed and they turned out to be so small that most cosmologists felt that it made sense to simply remove it from the equations, as Einstein had done. Other reasons for wanting a cosmological constant have come and gone. But today, we have good reason for believing that it is *not* zero. Current thinking is that Einstein did not make a blunder when he introduced it in his equations. First, let us take a closer look at the evidence for the Big Bang itself. After all, if the Universe is getting bigger then it must have had a definite moment of creation when it first started expanding.

Did the Big Bang really happen?

We are now very confident that our Universe was born about 15 billion years ago in a state of incredibly high temperature and density. What evidence do we have of this? The subfield of cosmology devoted to studying the birth of the Universe is known as cosmogony and is one of the most exciting areas of physics research. The most compelling evidence that our Universe was created in a Big Bang comes, of course, from the observation that it is expanding. As I have mentioned before, if the Universe is getting bigger, with the galaxies flying apart, then at some point in the past all the matter in the Universe must have been squeezed together.

Apart from the expansion of the Universe, the Big Bang model is also supported by two other crucial observations. The first is the observed abundances of the light elements. The fact that roughly three quarters of all the atoms in the Universe are hydrogen atoms and one quarter helium, the lightest and easiest elements to make, with only a tiny amount of all the other elements, requires a universe that was initially hot and dense but which rapidly cooled as it expanded. At the moment of the Big Bang, long before stars and galaxies had had a chance to form, all the matter in the Universe was squeezed together and there was no empty space. Immediately after the Big Bang (much less than a second), subatomic particles began to form and, as the Universe expanded and began to cool, these particles were able to stick together to make atoms. The conditions of temperature and pressure had to be just right for these atoms to form. If the temperature was too high the atoms would not have been able to remain intact. They would have been smashed apart in the hectic maelstrom of high speed particles and radiation. On the other hand, once the Universe had expanded a little, the temperature and pressure would have become too low to enable the atoms of hydrogen and helium to be squeezed together to form any other (heavier) elements. This is why mainly hydrogen and helium formed in the early Universe, a process that would have happened in the first five minutes after the Big Bang. Almost all the other elements had to wait until they could be cooked inside stars. The Big Bang model predicts the correct proportions of hydrogen and helium observed by astronomers.

The other piece of evidence in support of the Big Bang which, like the expansion of the Universe, was predicted theoretically before it was confirmed experimentally, is known as the cosmic background radiation. It is the 'afterglow' of the Big Bang explosion and is in the form of microwave radiation that permeates the whole of space and has a temperature today of about three degrees above absolute zero (or minus 270 degrees centigrade). To measure the temperature of this radiation experimentally, we do not need to stick a thermometer out in space. Instead, we use what look like giant satellite dishes which are called radio telescopes and which are so sensitive they can 'hear' this radiation's faint signal from deep space. This was done for the first time in the 1960s and has been confirmed many times since then with ever increasing sensitivity. If you find this hard to believe, I was impressed when someone informed me recently that we are even able to hear the hiss of faint radio waves given off by the planet Jupiter using a long-wave radio.

Today, there cannot be much doubt that the Big Bang did happen. There are other issues, however, that have yet to be resolved. Some are even being clarified at the time of writing this book. For instance, a few years ago we did not know whether gravity would one day halt the Universe's expansion and cause it to recollapse on itself, ending with all the matter rushing together in a cataclysmic implosion known as the Big Crunch, or whether the expansion would continue forever, with the Universe steadily getting colder and colder and ending in what is known as the heat death or the Big Freeze. Today we think we have the answer. It turns out that the fate of the Universe depends not only on how much matter it contains, but on the role of Einstein's cosmological constant. This makes cosmology a bit more complicated than we would have hoped. So I will wade through some of these big issues carefully, beginning with the shape of the Universe.

The edge of space

Consider the following two questions:

1. If the Universe is expanding but at the same time contains the whole of space, what does it expand into?
2. What is there beyond the edge of the Universe?

We feel as though there must be *something* beyond our Universe that can accommodate it as it expands. Believe it or not these questions are not purely philosophical or metaphysical. Science has an answer to both. It is just that we are not thinking about things in the right way. This is where all that stuff about higher dimensional geometry in Chapter 1 pays off. Naïvely, we think of the Big Bang as some explosion that *happened* at some point in time at a specific location in three-dimensional space. From this point, all of matter was ejected out and has been flying apart ever since. Wrong!

First of all, we have learnt that the Big Bang was not like a supernova explosion with all the matter flying away from a central point. The expansion of the Universe is a stretching of space itself, with the matter imbedded within space and carried along for the ride. Secondly, there is no point in the Universe where space explorers could travel to, planting a flag which states that: 'The Big Bang Happened Here'. Recall the example of the stretching sheet of rubber. The Big Bang happened everywhere on the sheet at once, and the stretching took place all over the sheet.

I don't expect you to feel happy about this just yet. Give me a couple more pages. I know I have not even answered the two questions yet. Let's try and tackle them head-on. Imagine that you are able to fly off in a rocket at very high speed and carry on going in a straight line—let's also assume that you are immortal and that the rocket has an unlimited supply of fuel. Would you ever reach a point beyond which you could not go? Some barrier beyond which there was complete nothingness?

According to Friedmann's model of the Universe based on Einstein's general theory of relativity (which we believe correctly describes the general features of the Universe), the answer is no, the Universe does not have an edge. There is no physical boundary that your rocket would eventually hit when it ran out of space. Nor would you ever reach a point beyond which there was nothing. If this abyss could be defined as space, then it is still part of the Universe, whether or not it contained any matter. So presumably, your rocket could just keep on going, and you would not have left the Universe, just entered an empty region of it.

Friedmann in fact found two different types of possible universe. If there is enough matter for gravity to one day halt the expansion and cause the Universe to recollapse (corresponding to the ball rolling back down a steep slope) then we would have something called a *closed* universe. If, on the other hand, there is not enough matter to halt the expansion then we would be living in an *open* universe[4].

Here is where I have to be careful. Friedmann's model makes an important assumption: that Einstein's cosmological constant is zero. This means that there is no force of antigravity acting at the moment to complicate things, even if it was responsible for setting the expansion going in the first place. The following discussion is therefore simplified[5] for the case of no cosmological constant.

---

[4] More correctly, he predicted three types of universe since a 'flat' universe would be one poised between open and closed.

[5] OK, by 'simplified' I mean compared with what the real Universe is probably like!

A closed universe

To visualize what a closed universe is like, we should go back to the example in Chapter 1 of the 2D'ers living on a spherical surface. Their universe is also closed, and is thus not infinite in size since the surface has a certain area. A sphere is said to have positive curvature, since if you were to move along two paths on the surface at right angles to each other, both would curve round in the same direction. Such a closed universe most certainly does not have an edge since the 2D'ers could travel anywhere they liked in the surface without reaching an edge. In fact, if a 2D'er were to head off in a rocket and travel in what is to him a straight line he would eventually come back to where he started from. That is exactly what would happen if we lived in a positively curved, closed universe; we would eventually come back to where we started from.

Remember also that for the 2D'ers, the inside of the sphere (and the outside) does not even exist. It is outside their two dimensions. If our Universe is closed then the simplest shape it can have is the surface of a four-dimensional ball known as a hypersphere. This is the equivalent of the 2D'ers' surface of a 3D ball, only it has one dimension more and is impossible for us to visualize. We should therefore spend a little more time thinking about the 2D'ers' universe since that is what ours would be like if we were to throw away one of our dimensions of space.

The following example is the standard way of explaining the concept of the Big Bang. Imagine the 2D'ers' universe is the surface of a balloon that is being blown up. The expansion of this universe is now exactly the same as the expanding flat sheet of rubber I discussed earlier. Every point on the surface of the balloon will be moving away from every other point. Now it becomes clear that the Big Bang is not *somewhere* on the surface of the balloon. It is more correct to think of it as the centre of the balloon itself, since not only is every point on the surface moving away from all others, they are all also moving away from the centre of the balloon. Even this picture is not quite right, however, since the interior of the balloon does not even need to exist. You see I have used the analogy of a balloon which is a 3D object, so that we can picture its 2D surface. After all, you would think that it makes no sense to talk about a sphere without imagining it containing an interior volume. But that is just for our own convenience. Such a closed 2D universe is able to exist *without* being imbedded in 3D space and we would say that its big bang happened everywhere on the surface at once, and since the whole surface was squeezed down to a point anyway, we do not need to specify where that point was situated within three-dimensional space. It is simply a handy way for our brains to visualize things.

To recap, if the Universe contains enough matter it will one day stop expanding and start collapsing. It would be a closed, finite universe which has positive curvature, and which will not have an edge just as the surface of a sphere does not have an edge. It might be helpful to think of it as expanding into a higher dimension, but this is really only an aid and the higher dimension does not necessarily have to really exist. In terms of where the Big Bang happened, we can say that it happened everywhere at once since the whole Universe would have grown out of a point and everywhere would have been confined to the same place. Whether that point was floating in higher dimensional space we do not know.

An open universe

The Universe is said to be open if it does not contain enough matter to stop it expanding[6]. In this case things get a little trickier to visualize. For a start, since this type of universe does not close in on itself, then the only way for it to avoid having an edge is for it to be infinite[7]. The simplest shape that such a universe could have would be for it to be flat, the three-dimensional analogue of the rubber sheet that would extend out infinitely in all directions. But for the Universe to have no curvature at all would be a very

[6] If you know a little more cosmology than you are letting on you might be aware that a universe can expand forever and still be closed. If you *are* a cosmologist, you don't need to read this book.

[7] Again, even this is not strictly necessary. Later on I will discuss how an open universe need not be infinite in extent.

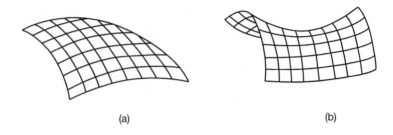

(a)   (b)

Figure 3.2. (a) Positive and (b) negative curved 2D space.

special case. It would be like the example of the ball rolling up the slope and managing to reach the top just as it runs out of puff and having no more energy to roll along the flat top. It is much more likely, if it is not going to roll back down again, that it would have some energy left over to keep going along the top. A universe corresponding to such a scenario will not be flat but curved. However, this time we say it has negative curvature.

So, by throwing away one of the dimensions of space we can make sense of the different types of curvature the Universe might have. If a positively curved universe corresponds, in a lower dimension, to the surface of a sphere and a flat universe corresponds to a flat two-dimensional sheet, what shape is a negatively curved two-dimensional surface? This one is not so easy. The correct mathematical name for such a shape is a *right-hyperboloid*, or hyperbolic, surface and is impossible to visualize properly. Very roughly, it has the shape of a saddle (see figure 3.2). The difference between the positive curvature of a sphere and the negative curvature of the saddle is that, whereas in the former any two paths on the surface that cross each other at right angles curve round in the same direction, such paths in the case of the saddle will curve in opposite directions. And the reason why the saddle is not an accurate depiction of a hyperbolic surface is that, as you move away from the centre of the saddle the surface gets flatter, whereas for a right-hyperboloid, the surface must have the same curvature everywhere. It is impossible to sketch such a surface.

Since the shape of an open universe is something very difficult to visualize, even in a lower dimension, let us see if we can do any better at understanding another puzzling feature. Namely, if the Universe is open and infinite then what does it expand into? By infinite I mean that space extends on forever in all directions. It would seem impossible that it can expand at all since all of space is used up and included *within* the Universe. Again, we can see the problem more clearly in two-dimensional space. For the case of a closed universe (the surface of the balloon), we can imagine the expansion to be outwards into a higher dimension, but for a flat sheet which has an infinite area, the expansion will always be in the plane of the sheet, and we cannot make use of the third dimension (off the sheet) as somewhere for it to expand into.

To solve this problem I need to explain a little mathematics. No one is comfortable thinking about infinity. I remember as a child being told that when we die we go to Heaven and we stay there for ever and ever. The thought of this use to depress me since I was uncomfortable thinking about something that would just go on and on without end however nice it was meant to be. Despite the difficulty most of us have contemplating the infinite, some mathematicians have made a living out of studying it. In fact, there are even different types of infinity.

Think of the sequence of integer numbers (or whole numbers): $1, 2, 3, 4, \ldots$ which goes on forever. We say that there is an infinite number of integers. But how about the sequence of even integers: $2, 4, 6, 8, \ldots$? Surely this sequence is also infinitely long. And since there are twice as many integers in total as there are even ones, we have two infinities with one seemingly twice as large as the other. What about the number of all numbers, not just the integers? For instance, let us consider the numbers:

$$0, 0.1, 0.2, 0.3, \ldots, 0.9, 1.0, 1.1, 1.2, 1.3, \ldots$$

and so on to infinity. This infinite sequence contains ten terms for every one in the sequence of integers. Is the infinity of terms in this sequence therefore ten times as big as the infinity of the integers? In mathematics, there is a whole subject devoted to the study of infinity. It turns out that the above three sequences all belong to the same class of infinity. But there are others. Consider the sequences of all numbers (called the set of real numbers) which includes all the fractions in between the integers. Even the interval between two consecutive integers such as 0 and 1 will contain an infinity of numbers $(0, \ldots, 0.00103, \ldots, 0.36252, \ldots, 0.9997, \ldots, 0.999999, \ldots, 1)$, since we can always think of a new fraction to slot in, however many decimal places it may have. There will likewise be an infinite number of fractions contained between 1 and 2, and between 2 and 3, and 763 and 764, and so on. Thus, we have a set containing an infinity of integers and an infinity of fractions between consecutive integers. This overall infinity is much 'stronger' than the infinity of integers, despite both being never ending! It turns out that there are in fact an infinite number of different infinities!

Where is all this leading us? The cosmologist Igor Novikov, considered by many to be Russia's answer to Stephen Hawking, uses the idea of different infinities to explain how an infinite universe is nevertheless still able to expand. Imagine that you check into Hotel Infinity, which has an infinite number of rooms— I've stayed in a few hotels that came close to this and I have certainly been lost in a few. You are told at the front desk that they are very busy that night and that there are already an infinite number of guests so all the rooms are occupied. You complain to the management that you had a reservation and insist that they find you a room for the night. "No problem" says the management, "in Hotel Infinity there is always room for more". They then proceed to move the person in room 1 into room 2, the person in room 2 into room 3, and so on, all the way to infinity. You are then given room 1.

What if an infinite number of guests arrive at once? Still no problem (forget for the moment about the infinitely long queue at the check-in desk). The management now move the person in room 1 into room 2, the person originally in room 2 into room 4, the person in room 3 into room 6, 4 to 8, and so on until all the guests are moved. Now all even numbered rooms are occupied. Since there is an infinite number of these rooms, all original guests are accommodated. This then leaves the infinity of odd numbered rooms now vacated and available for the new arrivals.

We can relate this example of the hotel guests to the space occupied by an infinite universe. It does not matter that new guests are arriving all the time. The hotel, being infinite, can always accommodate them. In the same way, an infinite space can always expand.

We now come to probably the most confusing feature of an infinite universe. If something is growing in size, then it would, by definition, take forever to become infinite. Thus, if our Universe is infinite in size today then it must also have been infinite in the past. In fact, it must already have been infinite in size at the moment of the Big Bang! This really flies in the face of the common notion of the Big Bang as the event when all of space was squeezed down to a point of zero size. This can at least be visualized for the case of a closed universe by dropping down a dimension and considering the example of the balloon. But an open universe was *never* zero-sized. The only way to think about this is to imagine that the Big Bang happened everywhere at once in an already infinite universe. Of course at any point in such an infinite universe the density would have been infinite too.

Another way to think of it is if the big bang of an open universe is like an infinitely long line. Even though it has an infinite number of points on it (since a point has zero extent) it still has zero volume over all. We could then imagine that our Visible Universe grew out from just one point (one big bang) of the line. I wouldn't push this analogy too hard though.

Finally, just to make sure you are totally confused, whatever shape the Universe has now, even if it is almost completely flat, it would have been *infinitely curved* at the Big Bang!

What shape is the Universe then?

Now that I have given you an idea of the different possible shapes our Universe might have (and probably a headache along the way), I will briefly review some of the recent discoveries and ideas in cosmology and what they tell us about the Universe. After all, if the Universe is going to one day collapse again in a Big Crunch

I think the public have a right to know. It may not be for another zillion years, but some people might just sleep more soundly if they were told.

As I have already mentioned, whether the Universe is open, closed or flat depends on how much matter it contains. This is, however, a bit of a problem if the Universe is infinite in size, because then it would also have to contain an infinite amount of matter, however thinly it was spread out! The reason for this is the cosmological principle which states that every part of the Universe is pretty much like every other and so the density of matter is roughly constant on the largest scale. This is a bit like saying that even though only one in every thousand rooms in Hotel Infinity is occupied, there would still be an infinite number of guests. So instead, physicists talk about the density of matter. This is the amount of matter per unit volume of space which should be a sensible number even if the overall volume is infinite.

If the Universe has a matter density that is more than a certain critical amount then the gravity of all this matter combined will be able to halt the expansion and cause it to recollapse. On the other hand, if the density is less than this critical amount then gravity can only slow down the expansion to a constant rate and never stop it. The Universe would be doomed to eternal expansion. Strangely enough, many cosmologists have good reason to believe that the density should be poised exactly at the critical value: balanced on a knife's edge between a universe that will one day collapse and one that will steadily expand forever. Instead, the density of matter would be such that its gravity will steadily slow the expansion rate down until it finally stops expanding. However, it would literally take forever for this to happen, so there would be no collapse. This corresponds to a flat universe, neither open nor closed. How did cosmologists come to believe that such an unlikely scenario is possible and why should they want it to be so?

The fact is that, as far as our telescopes can see, the Universe looks absolutely flat. It does not appear to have either positive or negative curvature. This was quite a problem for cosmologists since it was hard to believe that there would be precisely the right density of matter to keep space flat. If this were the case then gravity would always be applying the brakes on the expansion, continually slowing it down. This differs from a negatively curved open universe (with a density less than the critical amount) because in that case gravity would slow the expansion down only to some steady constant rate at which it would settle into forever.

Most cosmologists believe that the flatness problem has now been solved due to something called inflation. The simple explanation for the flatness of space that we observe would be if the Universe were much bigger than we think. In the same way, we do not observe the curvature of the Earth because we can only see a small part of the surface. The problem with this explanation is that the Universe does not appear to be old enough to have expanded to such a size. It is therefore thought that, when the Universe was just a fraction of a second old, it underwent a very short period of rapid expansion in which it grew to a size that was a trillion trillion trillion trillion times the size it was before. This number is unbelievably large and would be written as a one with 48 zeros! Thus, the Universe could have been curled up before the period of inflation. Then, in much less than a blink of an eye, it grew to such a size that we would never be able to detect any curvature, however far out in space we looked. This inflationary model of the Universe therefore requires the density to be very close to the critical value that would make it flat. In the mathematics, this density is denoted by the capital Greek letter *omega* (written as Ω). If the density is at the critical value, corresponding to flat space, we say *omega* has a value of one. If the Universe is positively curved and closed then *omega* is greater than one, and if it is negatively curved and open then *omega* is less than one.

We are not sure whether or not this rapid inflation of the very young Universe took place. Most cosmologists believe that it did, but the arguments for and against it are subtle and rely on a number of different issues, some of which have not been resolved yet.

Is it possible to measure the density of matter in the Universe directly? Cosmologists are confident it is. They rely on the cosmological principle which, if you remember, states that the Universe looks the same everywhere. In other words the density of matter everywhere is the same as it is in our little corner of the Universe. Of course by 'little corner' I mean the part of the Universe that we can see. So what *do* they see? It turns out that the density of matter visible to us (that is the number of galaxies in a given volume of space) is about one per cent of the critical value required for a flat universe. Oh, oh, we have a problem! Where could the other ninety nine per cent be?

Invisible matter

Part of this missing mass of the Universe is believed to be made up of something known enigmatically as *dark matter*. It is believed that there is probably between ten and forty times as much matter, or 'stuff', out there in space than we can actually see. This is not because it is so far away or hidden behind other objects, but because it is literally invisible. Here we go again, you're thinking, how can scientists even be sure about the things out in deep space that they *can* see, let alone stuff that is invisible! Well, yet again, the answer is surprisingly simple: galaxies appear to weigh a lot more than the sum of stars and other visible objects they contain and must also be made up of a cloud of invisible material that extends beyond the visible stars. This strange conclusion is reached from two quite independent routes.

The first is by studying the way stars on the outer rims of galaxies orbit the centre. If most of the mass of a galaxy is concentrated in its core, which is what one would expect since that is the region most densely populated with stars, then the outer stars should be revolving much more slowly than they actually are. The only way to explain the way these stars are observed to behave is if there is some kind of invisible form of matter, dubbed the dark matter halo, that surrounds, and extends further out from, the visible matter (the stars). This halo would have to contain many times more matter in it than all the visible forms of matter put together.

Another indication that galaxies are more massive than they appear is found by directly measuring their weight! This is done using Einstein's idea that the gravity of a massive object warps space around it. Remember from the last chapter that the general theory of relativity was first tested experimentally by observing the deflection of the light from distant stars as it passed close enough to the Sun's gravitational field. In the same way, a galaxy will deflect the light from a more distant galaxy when it is in the line of sight between the more distant one and us. The amount by which the light is bent tells us how much mass the nearer galaxy has. Again, we find that galaxies contain much more matter than just the visible stuff.

Until recently, it was not clear what this dark matter could comprise of. It was initially thought that it could be made up entirely of cold dead stars, black holes, planets, lumps of rock plus any other non-luminous (and therefore not visible) material that might be floating around out there that you thought was still down the back of your sofa. Such objects have been dubbed MACHOs, which stands for Massive Astronomical Compact Halo Objects. However, it turns out that there is a limit to how much of this kind of matter there could be, which is set by the proportion of the elements synthesized just after the Big Bang.

So a problem remains. We are now certain not only that dark matter exists but that most of it must be made up of a new kind of substance we have yet to discover.

Experiments carried out in Japan and announced in mid-1998 have suggested that part of it is probably made up of elementary particles called neutrinos. The trouble with these tiny entities, which were predicted theoretically in the early 1930s and discovered in laboratory experiments in 1956, was that no one had been sure they actually weighed anything at all. They are extremely elusive since they travel through solid matter, including our measuring devices, as though it was not there. In fact, billions of neutrinos, mainly produced in the Sun, are at this moment streaming through your body without you knowing it. Japanese scientists have now discovered that they do indeed have a very tiny mass which is enough to account for part of the invisible matter of galaxies, due to their sheer number. Even out in deep space it is estimated that there are, on average, several hundred of the little fellas in a volume the size of a thimble.

69

Even taking into account all the normal material in galaxies (whether visible or not) plus all the neutrinos, we still cannot account for all the mass that galaxies appear to have. Neutrinos would make up what is called 'hot dark matter' because they zip around at high speeds. We are now confident that there must be more, probably in the form of slower moving heavy particles that would make up 'cold dark matter'. The search is currently on in a number of laboratories around the world to find such new particles. My favourites are the WIMPs (weakly interacting massive particles) which might contribute many times more to the mass of the Universe than all the visible matter put together. Such particles have never been seen—well you wouldn't see them if they were invisible would you—but scientists can figure out what properties they must exhibit and are designing experiments to detect them.

So, the best current estimates for all the matter in the Universe (both visible and dark matter) only account for about a third of the density required to make *omega* equal to one and the Universe flat. It is looking increasingly likely that there is nothing more out there; that *omega* is in fact much less than one. How does this square with the theory of inflation which requires a flat universe? Are we going to have to modify it, or even abandon it altogether?

1998: a big year in cosmology

Measuring the expansion of the Universe is a tricky business. It involves a lot more than simply working out the speed that distant galaxies are receding from us by measuring the redshift in their light. First of all, it is hard to know for sure exactly how far away they are. And because they are so far away they tend to be, on average, younger galaxies—remember the light from them set off millions, even billions, of years ago—and younger galaxies tend to be bluer in colour and brighter because their stars are younger. On the other hand they are very dim because they are so far away. In addition to all this, galaxies come in all shapes and sizes and, while it is true that if enough of them are studied then we can reliably extract an average, all in all measuring the redshift of whole galaxies is not the best way of inferring the expansion rate.

There is a more reliable method. Remember from Chapter 2 that supernovae are so bright they briefly outshine the rest of their galaxy. In particular, type Ia supernovae (the result of a complete destruction of a star in a binary system after it has gained a critical mass by sucking matter from its partner) all shine with a certain brightness, or luminosity. They also flare up and die down within a certain time. This means they can be used as reliable standards for measuring distances. Recently, type Ia supernovae have been used to determine the rate at which the Universe is expanding; certainly the most exciting result in astronomy in 1998.

Detecting a supernova explosion of a star in a distant galaxy is extremely difficult because they are incredibly faint. What is even more incredible about the recent result is that very distant supernovae seem to be even fainter than they should be based on their distance. One reason for this could be because space is negatively curved (hyperbolic) which has the strange property of making distant objects faint because of the way their light spreads out in such a universe. But there is another more intriguing possibility. Maybe these supernovae are fainter because they are further away than we think. But that would mean that they should be receding faster than their measured redshift suggests. In other words they don't have a high enough redshift for their distance. Since the light reaching us from these supernovae set off when the Universe was much younger, their less-than-expected redshift indicates a slower expansion rate in the past! I know you may need to read this paragraph again to follow the logical order of arguments, but if the observations are correct then the bottom line is that the expansion of the Universe is NOT slowing down, but *speeding up!*

The only way for this to be possible is if a force of antigravity is driving the expansion, pushing the galaxies apart and stretching space. While gravity's influence gets weaker the further apart the galaxies are, antigravity gets stronger with distance, and so will drive the expansion even faster. The existence of this strange force is just another way of saying that the cosmological constant is not zero. But where does it come from?

The usual answer is that it is has to be due to some strange new form of invisible energy that is spread throughout the whole of space. This energy has the paradoxical effect of driving the expansion of space while at the same time contributing towards closing the Universe round on itself. That is, it would help to make up the missing fraction of *omega* to make it up to one, which is what many theoretical physicists would prefer. In fact, *omega* may even be a little more than one, making the Universe closed, even though it could expand forever. This makes the simple arguments based on Friedmann's model universe wrong. We can no longer say that an open universe is one that will expand for ever, while a closed one must one day collapse in a big crunch. The shape of the Universe and its destiny are no longer linked.

As for the origin of this energy of empty space, physicists are still working on it. It may be down to any one of a number of weird sounding Jargonese terms (which you may wish to impress your friends with) such as 'quantum fluctuations', 'phase transitions', 'topological defects' or, most wonderful of all, 'quintessence'.

Is the Universe infinite?

Here's another light-hearted topic you can discuss with family and friends instead of football[8]. If it turns out that the overall density of the Universe, due to all the visible and invisible matter and energy, is still not enough to make it closed, then conventional wisdom would suggest that it has to be infinite (to avoid having an edge you could fall off). Of course it may be that what appears to us to be an infinite flat universe may still be closed and just too huge for any curvature to ever be detected. In such a universe, the value of *omega* would be very close to one.

Most people, cosmologists included, would much rather not have to deal with an infinite universe. Over the past few years a new field of study called cosmic topology has emerged which is the study of the Universe's shape. One result of work in this

[8] Sorry about that. It's just that my two favourite topics of conversation are physics and football, and not necessarily in that order.

field that I was not aware of until recently was that even an open universe, whether flat or hyperbolic, can still have a finite size. In fact, and here is the fun part, even if the Universe is flat, it might turn out to be shaped like the higher dimensional equivalent of the surface of a doughnut (the ones with the hole in the middle). I know you must be thinking that the surface of a doughnut is hardly flat. But that is because it is only an approximation to the shape I am talking about. First of all, the surface of a doughnut is only two-dimensional. Secondly, even the 2D equivalent of the space I am referring to could not possibly exist imbedded in our 3D space. The correct name for such a shape is a Euclidean torus, and has the property, just like a doughnut's surface, of having more than one line joining any two points on it.

Of course if the Universe really is doughnut shaped then the missing mass is most likely sugar or cinnamon.

Why is it dark at night?

You might think that this is a rather trivial, even silly, question to ask. After all, even a child 'knows' that this is because the Sun sets below the horizon, and since there is nothing else in the sky anywhere near as bright as the Sun we have to make do with the feeble reflected light from the Moon and even more feeble light from the distant stars. Well, guess what? It's not as simple as that!

We have good reason to believe that even if the Universe is not infinite in size, it is probably so enormous that, for all intents and purposes, it is infinite. If so, then we come up against something known as Olbers' paradox. Simply stated, this says that the night sky has no right being dark at all. It should be even brighter than it normally gets during the day. In fact, the sky should be so bright, all the time, that it should not even matter whether the Sun is up in the sky or not.

Imagine you are standing in the middle of a very large forest. So large in fact that you can assume it is infinite in extent. Now try shooting an arrow in a particular (horizontal) direction such that it *does not* hit a tree trunk. In this idealized situation the arrow must be allowed to keep on going in a straight line without ever dipping down to the ground. You find, of course that it is impossible. Even if the arrow misses all the closer trees, it will eventually always hit one. Since the forest is infinite, there will always be a tree in the flight path of the arrow, however far away that tree is. It doesn't matter how dense the forest is either. If you were to chop down ninety per cent of all the trees, this would simply mean that the arrow will, on average, travel ten times as far before it encounters a tree trunk.

Now consider a simple model universe that is infinite, static (not expanding) and with stars evenly spread out. The light that reaches us from the stars is like the example of the arrow. It does not matter where we look in the sky, if the Universe is infinite we should always see a star in our line of sight. So there would not be any gaps in the sky where we do not see a star and the *whole* sky should be as bright as the surface of the Sun, all the time!

The real Universe may also be infinite, but in other respects it is not quite like the above simple model. First of all, the stars are not spread out evenly but clumped together in galaxies. This does not matter. It just means that the night sky should be as bright as an average galaxy, which is not quite as bright as the surface of an average star but still blinding. Secondly, our Universe is expanding. Does this make a difference? Physicists have carried out detailed calculations that have shown that this does not solve the problem, it just reduces it. So what is the answer?

It was thought that maybe space is filled with interstellar dust and gas that would block the light from the more distant galaxies. But if the Universe has been around for long enough, then even this material would slowly heat up, due to the light it has absorbed, and will eventually shine with the same brightness as the galaxies it obscures.

The true answer, the one which finally lays Olbers' paradox to rest, is that the Universe has not been around forever, so light from very distant galaxies has simply not had enough time to reach us. If the Big Bang happened 15 billion years ago, the galaxies that are further away than 15 billion lightyears (remember a lightyear is the distance covered by light in a year) are invisible to us because their light is still in transit and has yet to reach us. Admittedly, the discussion is complicated a little due to the expansion of the Universe, but what we can see in the sky is just a tiny fraction of the whole Universe. It is called the Visible Universe and we cannot, even with the most powerful telescopes, see beyond a certain horizon in space. Thus, the Visible Universe (our tiny corner of space) does have an edge even if the Universe as a whole does not.

Finally, we can turn Olbers' paradox on its head and say that *the real proof that the Big Bang happened is that it gets dark at night!*

Before the Big Bang?

One of the most popular questions asked by audiences when I lecture on cosmology concerns what there was *before* the Big Bang. After all, if the Big Bang really did happen 15 billion years ago, what then caused it to happen? What triggered the birth of our Universe in the first place? I will briefly state here three standard answers to the above. I will go through them here in reverse order of (personal) preference.

The first only applies if our Universe contains enough matter to eventually stop it expanding. In that case, it will one day in the very, very distant future begin to contract, ending finally in a Big Crunch. If this happens, and we think of the collapse into the Big Crunch as the time-reverse of the initial Big Bang, then the two events are equivalent. The Big Crunch of our Universe may therefore serve as a Big Bang for a new universe born out of the ashes of our own. And if this is the case, then our Universe may have followed an earlier one that had also expanded then collapsed. It may have been like this forever; an infinite number of universes, each expanding then collapsing in turn. Thus the answer to the question: what was there before the Big Bang? is that there was another universe, possibly similar to our own.

Since it now looks like the expansion of the Universe is speeding up, it will never be able to collapse again. Maybe the Big Bang was a one-off event. In that case, we must look to more exotic answers to the question. One which is gaining in popularity among the more mathematically inclined physicists is that the Universe was, until the Big Bang, part of a much grander space of ten (or eleven depending on who you talk to) dimensions. This universe is described as being 'unstable' as though it were unsure what to do with itself. The Big Bang came to the rescue causing it to 'quantum leap' into a more stable state. When this happened, six (or seven) of the dimensions curled up into an incredibly tiny ball leaving the three dimensions of space and one of time that we have today. This load of theoretical gobbledegook actually emerges naturally from the most sophisticated, yet at the same time most obscure, theories in modern physics, known as superstring theory and M-theory. Time will tell whether they are on the right track.

The final, and standard, answer is the following. If Einstein's general theory of relativity is correct, and we are confident that it is, then the Big Bang not only marked the birth of the Universe but the beginning of time itself. Asking questions about what happened *before* the Big Bang necessitates having time to imbed the word 'before' in. Since there simply was no time before the Big Bang, the question doesn't make sense.

Summary

I have covered a lot of ground in this chapter and it might be useful to briefly go back over some of the ideas I have discussed and state my level of confidence in their accuracy.

The Big Bang: Yup, almost sure to have happened. There are still a few physicists around holding out against it though. They argue that the Universe did not have a moment of creation but has been around forever. The theory they subscribe to is known as the steady state theory. What is interesting is that, despite so much evidence in support of the Big Bang, the steady state idea has yet to be satisfactorily laid to rest.

The expansion of the Universe: Like the Big Bang, there is no longer any real doubt about this.

The age of the Universe: Estimates at the end of the twentieth century put it at 15 billion years but this figure could yet be revised upwards (or downwards I suppose).

The shape of the Universe: Looks likely to be either open or flat, based on current estimates of the amount of matter it contains and the current rate of expansion. If I had to choose I would say that it is probably flat (or so nearly flat we would not be able to ever tell).

The size of the Universe: It is still possible for a flat or open Universe to be finite in size, though much larger than the furthest out we could ever see. Current theories would prefer it not to be infinite.

The fate of the Universe: Regardless of what shape or size the Universe might have, the latest results measuring the expansion rate from distant supernovae suggest strongly that the Universe will expand forever ending in a Big Freeze. In a way, this is easier for many people to cope with, for at least then time will go on forever. It is one thing to talk about the Big Bang being the beginning of time, but the Big Crunch would mark the very end of time. Not only would nothing survive after it, but the word 'after' would have no meaning!

Inflation: This theory is looking quite healthy at the moment, although there are a number of different versions of it. Most require the Universe to be flat, but a new theory called 'open inflation' is currently being developed. This does away with the requirement for flatness and allows for the bizarre concept of an infinite, open universe to fit inside a finite volume 'bubble' which floats in some external space[9].

Antigravity: At the moment the cosmological constant is back in fashion, suggesting that there is a repulsive force of antigravity pushing matter apart which is driving the expansion of the Universe. But we still do not understand its origin.

---

[9] Don't blame me, I didn't think this up!

4

BLACK HOLES

More to light than meets the eye!

Light is strange stuff indeed. Unless you have a scientific background, chances are you have not wasted too much time wondering what light is made of. Surely, you might think, it is what emanates from objects like the Sun, electric light bulbs, torches, candles, fires and so on and, whatever it is made of, enters our eyes and we 'see' things. When light bounces off an object, it carries with it into our retinas information about the shape and colour of that object. But what is light itself ultimately made of?

I have already described how the light from an object that is moving away from us becomes redder due to a stretching of the light's wavelength. This implies that it is not made up of physical material that we can touch. In fact, we are taught at school that it is just periodic, oscillating waves of energy, like sound waves or the ripples on the surface of a pond when a stone is thrown in. All the experiments you would have done in school science labs would have probably confirmed this. Light waves reflect off mirrors, get focused through lenses and are split up into the colours of the rainbow, known as the visible spectrum of light (when sunlight is passed through a prism).

Some of these experiments with light can be a lot of fun, and I recall enjoying building a box camera as a child and trying to understand how rays of light passing through a tiny pinhole in the side of the box could still produce the (upside down) image on the back. Fortunately for those with a lower boredom threshold than your average physicist, it turns out that light is nowhere nearly as boring and straightforward as we are led to believe at school. In fact it is so weird that all science teachers sign a secret document in which they vow never to divulge the true nature of light to the innocent and unsuspecting children. By the time the children grow up they will either be totally disinterested or would simply refuse to believe that something as familiar to our everyday life as light could hide so much mystery and yet be so fundamental to the workings of the Universe.

OK, I admit that sinister secret covenants signed by school teachers sounds like something out of a Roald Dahl story. Of course there is no global conspiracy to hide the true nature of light, but I am serious when I say that there is more to light than meets the eye!

Sound is a simple example of a wave. An object is said to make a sound if it sets the air molecules around it vibrating. These molecules collide with others nearby setting them in motion and so on all the way into our ears. The air molecules inside the ear then set the ear drum vibrating and our brains translate this vibration into something we know as sound. But at no point can we say that a material 'substance' has travelled from the object making the sound to our ears.

Light is much more than this. In Chapter 6 I will reveal how light is fundamental to the very nature of space and time. The physicist David Bohm summed it up when he said that "when we come to light, we are coming to the fundamental activity in which existence has its ground". For now, and for the purpose of discussing black holes, let us investigate what light actually comprises.

Isaac Newton firmly believed, based on his famous experiments with prisms, that light was composed of a stream of tiny particles he called corpuscles. This, he claimed, was obvious since light most certainly did not behave like sound waves. Light rays always travelled in straight lines (the bending of light due to gravity was a long way from being discovered) and cast sharp shadows. Sound waves worked their way around obstacles and could easily bend round corners. Sound waves need a medium to travel through; a material substance made up of atoms which, by oscillating, carry the energy and frequency of the sound. This is why the posters advertising the cult film *Alien* carry the, quite correct, caption 'In space, nobody can hear you scream', since in space there is no air to carry the sound waves. Light, on the other hand, is not at all like this and clearly has no difficulty travelling through empty space.

For these reasons Newton was convinced that the particle theory of light was correct. But not everyone was convinced, and it took over a century for clear proof to emerge that Newton's theory was not the whole story. At the beginning of the nineteenth century Thomas Young discovered that the reason light did not appear to bend round corners was because the effect was so small. The wavelength of light is so short compared with that of sound that the amount of bending, called diffraction, is hard to detect. Nevertheless, Young achieved this by sending light through very narrow slits and showed that, when it hit a screen on the other side, it formed a row of light and dark fringes in a way that would be impossible to explain if light were composed of particles. Such interference fringes, as they are known today, are explained in every physics textbook as being due to the way the peaks and troughs in the light waves from the two slits reinforce and cancel. We have all observed this effect at school, with varying degrees of excitement, in ripple tank experiments.

So was Newton wrong? Is light a wave rather than a stream of tiny particles after all? In the late nineteenth century it appeared that Young's interpretation of light as a wave was put beyond any possible doubt when the Scottish physicist James Clerk Maxwell developed a set of equations which showed that all the light we see is a form of electromagnetic radiation, which we know now includes other forms such as radio waves, microwaves and x-rays, as well as infrared and ultraviolet radiation. Light, it turned out, was made up of a combination of electric and magnetic fields, vibrating at right angles to each other, that could travel through empty space. So light was a wave after all. Was this the end of the story?

Far from it. Enter Albert Einstein, who won his Nobel prize in physics due to a paper he wrote in 1905—which, amazingly, had nothing to do with his theories of relativity. The paper explained something called the photoelectric effect and proved that Newton was not entirely wrong after all. Light at its most fundamental level is made up of tiny entities called *photons*.

So what of Young's interference fringes? And what of Maxwell's electromagnetic waves? What on earth is going on here? Surely light must make up its mind what it is made of: waves or particles?

There have been numerous books explaining what is going on. It turns out that light is indeed schizophrenic. Sometimes we see it behave like a periodic wave and other times like a stream of particles. It depends on what type of experiment we do! If you don't like this, then tough. I told you light was weird. The theory that describes the rules for the behaviour of light is known as QED, which stands for quantum electrodynamics, and was developed by, among others, the American physicist Richard Feynman in the late 1940s. QED, as its name suggests, is itself derived from a much broader theory in modern physics called *quantum mechanics* which describes the behaviour of not just light but all matter and energy at its most fundamental level (the level of atoms and smaller).

Quantum mechanics was developed in the mid-1920s by a number of European physicists, including Einstein. It describes things like how a single atom can be in two different places at the same time, and how tiny particles can spontaneously appear out of nowhere then quickly disappear again. The world's top physicists all agree that if anyone is not uneasy with what quantum mechanics tells us about the world we live in then he or she has probably not really understood quantum mechanics. Despite this it has been the single most successful and important scientific discovery of the twentieth century. Quantum mechanics underpins the whole of modern chemistry and the whole of modern electronics. Without it we would not have been able to understand the structure of crystals, or invent the laser or the silicon chip. Without an understanding of the rules of quantum mechanics there would be no televisions, computers, microwaves, CD players, digital watches and so much more that we take for granted in our technological age.

We will leave our discussion of light, for now, with a quote from Einstein from 1951 (four years before he died):

> *"All these fifty years of conscious brooding have brought me no nearer to the answer to the question 'What are light quanta [photons]?' Nowadays every Tom, Dick, and Harry thinks he knows it, but he is mistaken."*

Having given you this brief introduction to the nature of light I can begin to discuss the properties of black holes.

Invisible stars

To begin our story of black holes we must go back two hundred years to the end of the eighteenth century, since that is when scientists first realized that black holes might exist. Back then they were known as invisible stars and their existence followed logically and reasonably from a combination of Newton's law of gravity and his particle theory of light.

Until fairly recently it was thought that the first person to predict the existence of black holes was the world famous French mathematician and astronomer Pierre Laplace in 1795. It is now clear that he was beaten to it by an English geologist named John Michell, who was rector of Thornhill Church in Yorkshire.

Michell is considered to be the father of the field of seismology and was the first to explain, in the aftermath of the Lisbon earthquake of 1755, that quakes started as a result of the build-up of gas pressure from boiling water due to volcanic heat. He also pointed out that earthquakes could start underneath the ocean bed and that the Lisbon one was an example of this. His ideas on the formation of black holes in space were presented to the Royal Society of London in 1783. Both Michell and Laplace had based their quite similar arguments on the idea of escape velocity.

I recently watched a documentary on television about amateur rocket builders. These guys take their hobby very seriously and have competitions to see which rocket can reach the highest altitude before Earth's gravity reclaims it. The problem is, of course, that rockets need to achieve an escape velocity before they can break away from Earth's pull and make it into outer space. Surprisingly, space is not really that far away—about a two hour drive by car if we were able to head straight up. It is just that whatever speed an object starts off at, gravity will immediately begin to slow it down, and so it has to start off fast enough to allow for this slowing down. Remember that the force of gravity becomes weaker with distance and so a rocket does not need to be travelling very fast once it has reached a certain height. In practice, rockets only get into orbit gradually, by firing engines in successive stages.

The escape velocity on the surface of the Earth is eleven kilometres per second (or forty *thousand* kilometres per hour). On the moon it is a little over two kilometres per second, which is why the Apollo missions' Lunar Modules did not need such large rocket engines to leave the moon and return to Earth.

The escape velocity on the Sun is 620 kilometres per second. This is a number which Michell had worked out, based on the size and density of the Sun. He also knew with some accuracy another figure: the speed of light, which had been measured a century earlier, and which is 500 times bigger than the Sun's escape velocity. Michell therefore calculated that a star 500 times bigger than the Sun, but with the same density, would have an escape velocity equal to the speed of light.[1]

Michell was following the conventional wisdom of the time; that of Isaac Newton, and believed that light was composed of particles. It therefore followed that light should be affected by gravity like any other object. But a star with the same density as the Sun, but more than 500 times as large, would have an escape

---

[1] Notice that I am using both the terms velocity and speed in the same sentence. They mean the same thing here and it is purely a matter of convention that both are used. There is a technical difference between the two but it need not concern us.

velocity that exceeded the speed of light, and so the particles of light would not be fast enough to escape its gravitational pull. Such a star must therefore look black to the outside world. In fact, it would be invisible!

Michell had explained the 'black' part but we have to fast forward to the twentieth century to understand what the 'hole' bit means. After all, an invisible star, as explained by Michell and Laplace, is not really very interesting. In fact, using the idea of escape velocity to explain black holes is a bit like saying that the Big Bang was just a very big explosion of matter and energy, and leave it at that. As I explained in the previous chapter, the Big Bang didn't happen somewhere in space and at some point in time but rather encompassed space and time within it in a way that is extremely hard for us to grasp. In the same way, black holes are much more than large dense clumps of dead star whose gravity is too strong to let light escape. In fact they differ from Michell's invisible stars in some astonishing ways. For a start, Michell's black stars are solid objects of some definite size. Black holes, as we understand them today, comprise almost entirely empty space! In fact they are literally holes in space, inside which the properties of space and time are completely altered. And although we have never come face to face with a black hole, we have a rough idea what it would be like to fall into one (not very nice). The reason for this confidence is our trust in Einstein's general theory of relativity. For if general relativity is correct, and we have no reason to doubt it so far, then it suggests that black holes not only exist in our Universe but are an inevitable consequence of Einstein's version of gravity. One of the world's leading experts on general relativity, Kip Thorne, goes so far as to state that "the laws of modern physics virtually demand that black holes exist".

We broke off at the end of Chapter 2 after briefly describing what happens to a star much more massive than the Sun when it runs out of its nuclear fuel. Having completed our bumpy—but I hope exhilarating—ride through some of the ideas in cosmology in the last chapter we are now ready to look in more detail at exactly how and why a black hole forms. Remember that Einstein's view of gravity is to do with the curvature of space, and the stronger the gravitational field of a massive body, the more curved and distorted the space will be around it.

When a large star explodes as a supernova, it will often shed most of its mass into space leaving behind a neutron star core which no longer has enough mass to collapse any further. Inside such a dense object, matter is packed so tightly that even atoms cannot retain their original identity. In normal matter, such as the stuff that makes up everything around us, including ourselves, atoms are mostly empty space themselves despite being so small. They comprise of a tiny core known as the atomic nucleus which is surrounded by even tinier electrons. The laws of quantum mechanics govern how these electrons behave within atoms and explain why they keep their distance from the nucleus. Inside a neutron star, gravity is so strong that the atoms get squashed together and the electrons are squeezed into the nuclei. The laws of quantum mechanics state that there will now be an outward pressure that prevents the neutron star from collapsing any further under its own weight.

What if, after a star has shed part of its mass in a supernova, its remaining core is still above some critical mass (roughly three times the mass of the Sun)? Now even a highly compact object with the density of a neutron star is not 'solid' enough. Its matter does not have sufficient internal pressure to withstand further gravitational contraction. In fact the star has no choice but to keep collapsing. Rather than slowing down, the gravitational collapse actually speeds up. It is rather like a ball rolling over the crest of a hill. Once it gets past the highest point and starts to roll down the other side it will just get faster and faster. The question is, what happens next? Surely the collapse must stop somewhere? The star is being squeezed smaller and smaller with the matter inside it being packed more and more closely together.

We now see that the escape velocity from the surface of a star depends both on its mass and its size. So we do not need a star that has the same density as the Sun and which is five hundred times bigger for it to have an escape velocity exceeding the speed of light. We can achieve the same result if the Sun itself could be squeezed down to a size just a few kilometres across since then, despite it having the same mass (the same original amount of material) as it did before being squeezed, it is now much more densely packed. Such a density would be considerably greater even than that of a neutron star (which would typically have an escape velocity of about half the speed of light) and will continue to collapse further to form a black hole.

Thus Michell and Laplace's argument that a collapsing star would eventually disappear from view should also hold for collapsing stars that have a density that is greater than the critical value for a neutron star. But this does not even begin to describe the exotic nature of black holes. After all, we would like to know what, if anything, can halt the apparent runaway gravitational collapse of such an object, even if we can no longer 'see' what is happening.

The clue is in the fact that Michell and Laplace were using Newton's version of gravity and not Einstein's. In Chapter 2 it seemed as though the main difference between Newtonian and Einsteinian gravity was in the way it was interpreted. Newton described it simply as an attractive force between any two objects, while Einstein said that it was a curving, or warping, of space around an object due to its mass, which causes other objects close by to roll into the dent in space around it and thus move closer to it. But surely the final result is the same however we choose to interpret it? It turns out that this is not the case. Once gravity becomes very strong (such as in the vicinity of a collapsing massive star) Einstein's version of gravity begins to depart radically from Newton's. In fact, Newtonian gravity is said to be only approximately correct. It works well in the weak gravitational field of the Earth, but to understand black holes we must ditch it completely.

As soon as Einstein completed his general theory of relativity in 1915 he began trying to *solve* his field equations. These equations were the complicated (yet mathematically beautiful) embodiment of his ideas on the connection between matter, space and time. But being able to write down the equations is only half the battle. They then have to be applied to particular situations and scenarios which involves much more than simply 'putting' numbers into a formula, but involves page after page of tedious and complicated algebra. This is in sharp contrast with the mathematics of Newtonian gravity, which is so simple it is taught at school. The first exact solution of the equations of general relativity was obtained by a German astronomer named Karl Schwarzschild. He completed his calculations while on his death-bed, having contracted an incurable skin disease during the First World War, and only a few months after Einstein had himself published his work. The Schwarzschild solution, as it is now known, described the properties of space and time due to the gravitational field around any spherical concentration of mass. It was only later realized that Schwarzschild's result contained a description of a black hole in space. In fact, it was not until 1967 that the American physicist John Wheeler first coined the phrase 'black hole' which has since captured the public's imagination so spectacularly.

Beyond the horizon

Schwarzschild's solution of Einstein's equations states that when a massive enough body collapses under its own weight it will reach a critical size beyond which there is nothing to stop it from collapsing further, no matter how squashed the body is. This is the size which, according to Newtonian gravity and as calculated by Michell, the star must be for the escape velocity to equal the speed of light. But there is a major difference in relativity.

If we use only the rules of Newtonian gravity then, provided the internal pressure of the collapsed star is strong enough, there is no reason why it could not stop collapsing at, or just beyond, this critical size. It just depends on the stage at which the molecules, atoms or even the subatomic particles say enough is enough, we will not tolerate any further compression.

The force of gravity according to Newton grows in what is known mathematically as an inverse square relation with distance. This means that if a star collapses to a size that has a radius that is half its original value then the force of gravity on its surface will be four times as strong. If it collapses to one third of its original radius then gravity will be nine times as strong, one quarter and it is sixteen times as strong, and so on. As the radius gets smaller the force of gravity gets bigger. If it were possible for all the mass of the star to be squeezed to a pinpoint of zero size (it now has a radius of zero) then the force of gravity would be infinite.

General relativity says something dramatically different. If a star were to collapse down to some critical size, such that its escape velocity equals the speed of light, then *the gravitational force on its surface would be infinite!* By this I mean the force that would be required to stop it from collapsing further would be infinite. The radius of this critical size is called the Schwarzschild radius and marks the boundary of a black hole. Now we see that the collapse *must* continue beyond this radius. If you are mathematically inclined[2] you may be wondering[3] how it is possible for this force to be infinite at the Schwarzschild radius and grow even stronger after the star has collapsed further. How can anything get bigger than infinite?! The answer lies in the discussion (in Chapter 2) of objects in free fall. Recall that when you are in free fall, say at the end of a bungy jump rope (and before you reach the bottom) your acceleration cancels out the effect of gravity and you do not feel any gravitational pull. In the same way, as the surface of the star collapses through its critical radius it is in free fall, and its surface does not feel the gravitational pull of the interior of the star. This is why the star cannot stop at the critical radius since it is impossible now to stop it from collapsing further.

Within the Schwarzschild radius, nothing—not just light—can escape. Imagine a sphere that has a radius equal to the Schwarzschild radius and which surrounds the collapsed star. Such an imaginary spherical surface is known as the event horizon and is an artificial boundary in space which marks the point of no return. Outside the horizon gravity is strong but finite and it is possible for objects to escape its pull. But once within the horizon, an object would need to travel faster than light to escape, and this is

[2] Don't worry if you are not.
[3] Don't worry if you are mathematically inclined and not wondering!

not allowed. Thus the event horizon is a rather unpleasant concept in that it allows one-way traffic only.

The event horizon is an appropriate name since it can be compared (loosely) with the common meaning of 'horizon' on Earth. This is the artificial line that marks the furthest distance we can see and appears as the place where the ground meets the sky. We understand that this boundary is due to the curvature of the Earth and, since light travels more or less in a straight line near the surface of the Earth, we are unable to see beyond it. In the same way, a black hole's horizon marks the boundary beyond which we cannot see any 'event'. But unlike the horizon on Earth which continually moves back as we approach it, an event horizon is fixed and we can get as close to it as we like, and even pass beyond it if we were foolish enough.

All bodies have their own potential event horizon with its own Schwarzschild radius. Even the Earth could be made into a black hole, but since it does not have enough mass to collapse by itself it would have to be squeezed from the outside. Don't ask me how, I am just saying that if it *could* be squeezed hard enough then it would eventually pass through its own event horizon, by which time its collapse would be self-sustaining. The Earth's Schwarzschild radius is less than half a centimetre, which means that any black hole with as much matter in it as our planet would be the size of a pea.

Once a collapsing star has contracted through its event horizon, nothing can stop it from continuing to collapse further until its entire mass is crushed down to a single point. This is called the *singularity* and is a very strange entity indeed. It is so strange in fact, that the laws of physics that work—as far as we know—perfectly well everywhere else, describing the behaviour of the tiniest subatomic particle to the properties of the whole Universe, break down at a black hole's singularity. It is therefore quite a relief for the outside Universe that the event horizon shields us from such a monstrosity.

Without an event horizon, who knows how a singularity would corrupt the laws of physics outside the black hole. In fact, the horizon is so important that physicists have invented the grand-sounding law of cosmic censorship which, they hope, applies everywhere in nature. They act like the Mary Whitehouses[4] of cosmology guarding the Universe against the chaos, unpredictability and infinities of the singularity. So what does this law state? Quite simply that: "Thou shalt not have naked singularities". You have to bear in mind that this rather tongue-in-cheek statement is really only a hypothesis and may well turn out, in certain theoretical scenarios at least, not to hold. For instance it is claimed that tiny black holes, smaller than atoms, may have been created just after the Big Bang and slowly evaporate away through a process known as Hawking radiation (which we will meet later in the chapter). Some calculations have shown that what might be left behind at the end of this evaporation are naked singularities. However, this is by no means clear.

According to the equations of general relativity the singularity is the place where matter has an infinite density, space is infinitely curved and time comes to an end. There is a common misconception that time comes to an end at the event horizon. This is because of what distant observers see as they watch something falling into a black hole. I will deal with this later; for now I want to return to the singularity marking the end of time.

Ring any bells? It should do. This is precisely how I described the Big Bang itself. Only then the Big Bang marked the beginning rather than the end of time. Apart from that the two cases are remarkably similar with the Big Bang being the mother of all singularities; a naked one to boot.

Back to black holes and their interior which, as defined by the event horizon, is completely empty space apart from the singularity in its centre (and apart from any matter that has been captured by the hole and is falling in). The reason why the singularity has infinite density can be seen if we consider how we calculate the density of an object. It is the ratio of its mass to its size. Thus if an object of any mass has zero size then to

[4] Mary Whitehouse is honorary secretary of the National Viewers' and Listeners' Association in the UK and has campaigned for many years to 'clean up' these and other media by restoring a more 'balanced' view of sex in programmes for family viewing.

obtain its density we must divide a number that is non-zero by zero. And this, believe me, is a highly undesirable thing to do in mathematics. Try it for yourself. Divide any number by zero on a pocket calculator. Mine gives me the symbol '-E-' which stands for 'error' since a humble calculator cannot cope with infinity. Come to think of it neither can my powerful workstation computer I use for research at university. If it encounters a division by zero the program it is running simply crashes. At least it does me the courtesy of telling me where the problem is in the code. It turns out, however, that the singularity is not quite as nasty as this. When we apply the rules of quantum mechanics, as we must do at this level, we discover that the singularity has an extremely tiny (much smaller than an atom) but *non-zero* size. Many of the details of the physics have yet to be ironed out, since applying the rules of quantum mechanics at the same time as the rules of general relativity is something no one knows how to do properly yet.

A black hole is therefore very simple in its structure. It has a centre (the singularity) and a surface (the event horizon). All else is gravity. Of course what makes black holes so fascinating is the way their tremendous gravity affects space (and time[5]) nearby.

A hole that can never be filled

So far, I have described the formation of a black hole in terms of gravitational collapse. But we have learnt that Einstein's view of gravity is in terms of the curvature of space. A black hole can also be described in this way. Think of the example I used in Chapter 2

[5] Aficionados of relativity theory and black holes may be wondering by now why I have steered clear of the discussion of time and how it is affected by gravity according to Einstein. I am aware that the traditional way of teaching black hole physics is within a unified description of the curvature of space and time. Indeed it will be very difficult for me to describe what happens if we were to fall into a black hole without discussing how our perception of time is changed. However the whole concept of time has been so revolutionized by Einstein's work that it deserves a more careful and gentle introduction for non-physicists. I will therefore, as much as possible, postpone any discussions about the nature of time, inside and outside black holes, till later in the book.

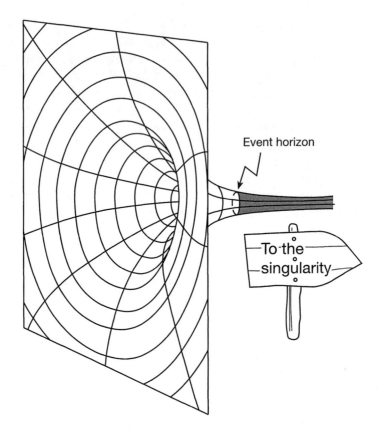

Figure 4.1. A black hole in 2D space.

of placing a heavy object in the middle of a rubber sheet causing it to sag under the weight. This dip in the sheet is equivalent to the curvature of space under the influence of a massive body. If the body is much heavier, the dip would be deeper. A black hole corresponds to the case when a very heavy, yet point-sized, object causes the rubber (space) to be curved and stretched down into an infinitely deep cone-shaped hole (figure 4.1). The event horizon here corresponds to a circle somewhere inside the rim of this bottomless pit beyond which there would be no escape.

There are two interesting observations we can make here based on this simple picture. Firstly, a black hole can never be plugged up or filled in with matter. The more matter that is poured into a black hole the bigger it gets. It will feed and grow.

Secondly, the size of a black hole, as measured by the volume within its event horizon, is really only a measure as seen by an outside observer. As an example, a black hole formed by the collapse of a star ten times the mass of the Sun will have a Schwarzschild radius of thirty kilometres, making the black hole roughly the size of a large city. Of course, an outside observer cannot see beyond the event horizon anyway and therefore can have no idea what things are like inside. But if space inside the horizon forms an infinitely deep hole then the distance from the horizon to the singularity should really also be infinite. In reality, and as I shall describe later, if you were to fall into a black hole then it would take you only a very short time to reach the singularity since space and time go haywire inside the horizon and one cannot use simple rules such as speed equalling distance divided by time.

From the outside, all black holes of the same mass look identical; we are unable to learn anything about the object that created the black hole in the first place, even being ignorant of its original chemical composition. All that information has been lost from our Universe forever. William J Kaufmann makes this point clearly in his excellent book, *Universe*. He considers two hypothetical black holes—one produced by the gravitational collapse of ten solar masses of iron and another from ten solar masses of peanut butter. Once they have both collapsed beyond their event horizons, they become identical and we are unable to tell which black hole was formed from which substance.

A common misconception regarding black holes is that they will eventually gobble up everything in the Universe. This is not true. Gravity is said to behave relativistically in a region where the predictions of Einstein's version depart radically from those of Newtonian gravity. For example, a black hole with a Schwarzschild radius of thirty kilometres will only cause the gravitational field around it to behave relativistically out to a distance of a thousand kilometres. Outside this range, the black hole obeys the rather boring laws of Newtonian gravity and behaves like any normal star of that mass in the way it affects the motion of distant objects.

Spinning black holes

So far I have restricted myself to discussing the simplest kind of black hole: one that is described by the Schwarzschild solution of Einstein's field equations. This is really only an idealized scenario. A real black hole would be spinning too. We know that stars spin about their axis in the same way that the Earth does. Therefore when they collapse they will spin even faster. Let us examine briefly why this should be so.

An important quantity in physics is known as angular momentum and is possessed by all rotating objects. The reason it is so important is that it is one of those quantities, like energy, that is said to be *conserved*, which means it stays the same provided the rotating object is not subjected to an external force. Angular momentum depends on the mass of the object, the rate it is spinning and its shape. Think of an ice skater spinning with her arms extended outwards. As she brings her arms closer to her body and folds them against her chest, she will spin faster. The reason for this is that her angular momentum must remain constant— ignoring friction of the blades on the ice—and, by bringing her arms in, she has altered her shape which would reduce her angular momentum if this were all that happens. However, she will also spin faster to compensate for this and keep her angular momentum the same. This increase in the rate at which she spins is not something that she does deliberately; it happens automatically. Aren't the laws of physics clever? A collapsing spinning star behaves in the same way: its reduced size must be countered by it spinning faster in order to maintain its angular momentum. This is why pulsars (the spinning neutron stars we met in Chapter 2) spin so rapidly.

According to Newton's version of gravity we cannot tell the difference between the gravitational effects of a spinning spherical object and one that is not spinning (as long as it does not wobble about as it spins). Here again, general relativity is different. A spinning black hole literally drags space around it to form a gravitational vortex, rather like the way water circles round a plughole. In such a region of space an orbiting body would have to accelerate in the opposite direction to the spin of the black hole just to stand still! This strange result provides us with a means of measuring the rate at which a black hole is spinning which, along with its mass (from which we can deduce its size), is the only other quantity there is to describe all we can about a black hole[6]. To measure the spin of a black hole we need to put two satellites into orbit in opposite directions around it. Since the satellite that is orbiting in the opposite direction to the black hole's spin must move 'against the tide' of moving space, it will take longer to complete one full orbit since by covering more space it has travelled further. The difference in orbit times between the satellites tells us the rate of spin.

This region where space is dragged round a spinning black hole is called the ergosphere. It means that a spinning black hole will have two horizons: an inner, spherical, one which is the original event horizon and from which nothing can escape and an outer, bulged-out at the equator one which marks the surface of the ergosphere (figure 4.2). Within the ergosphere, the dragging is so strong that nothing can stand still. However, an object that falls into the ergosphere can still escape again, as long as it does not stray within the event horizon.

Falling into a black hole

One of the most fascinating things about black holes is what happens to objects/suicidal astronauts that fall into them, and how this compares with the way things look to an observer watching from a safe distance. Let us first consider what it would be like if you were unlucky enough to fall in to one.

---

[6] We can also measure the electric charge of a black hole but this would be very small and is only of interest theoretically. In practice, a charged black hole will always eventually be neutralized by sucking in particles which have the opposite electric charge to it.

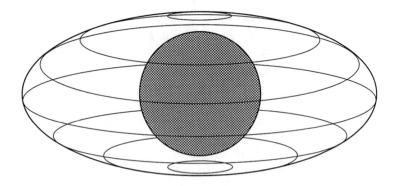

Figure 4.2. The ergosphere surrounding the event horizon of a spinning black hole.

One aspect of gravity that has not been mentioned so far is the tidal force. We know that the gravitational pull of a body becomes weaker the further away we are from it. Surely it then follows that, just by standing on the ground, your feet should feel a stronger pull due to the Earth's gravity than your head, which is further away from the surface. This is in fact true, but the difference in the gravitational field of the Earth is so tiny over such a small distance that you would never feel it. We can, on the other hand, clearly see the tidal effects of the Moon's gravity on the Earth. This is because the side of the Earth facing the moon feels a stronger gravitational pull than the opposite side which gives rise, as the Earth spins, to the daily tides of the seas from which the tidal force derives its name.

When it comes to black holes the gravitational force is changing much more dramatically and you are able to feel the tidal effect even along the length of your body. This becomes unbearably strong and will ultimately rip you to shreds long before you are finally crushed at the singularity.

A small black hole, of the order of several solar masses, has tidal forces so extreme that any astronaut venturing too close would be killed long before he or she has even crossed the event horizon! Not very nice is it? You would think that you might at least be given the chance to get close to the horizon without too much trouble. Luckily we have good reason to believe that there exist black holes with masses millions of times that of the Sun. Such supermassive black holes have much gentler tidal forces and one could easily cross the event horizon of such a hole without feeling any discomfort. As you continue to free fall towards the singularity the tidal forces will gradually grow in intensity. Thus although you will eventually be ripped apart then crushed to a point of infinite density at least you can now do a little sightseeing on your way down.

Throughout this book you might have gathered[7] that I have been trying to postpone the discussion of gravity's weird effect on time until the next section. I cannot, however, do black holes justice without relaxing my resolve on this a little. Inside a black hole space and time are so warped that the distance from the event horizon to the singularity is not a distance in space in the normal sense (in the sense that it can be measured in kilometres or some other appropriate unit of length). Instead it becomes a time direction. Basically the radial distance to the centre of the hole is interchanged with the time axis! Just a minute, you think, we have been discussing the size of black holes in terms of their Schwarzschild radius which is most definitely measured in units of length. The difference is that the Schwarzschild radius is a radius *as viewed from outside the hole*. Imagine observing a black hole against a bright backdrop that would clearly highlight its dark horizon, such as a luminous gas nebula. The distance across this black disc is its diameter, or twice its Schwarzschild radius. Once inside the black hole things are very different.

This interchange of space and time explains why any object falling into a black hole has no choice but to move inwards towards the singularity. Physicists liken this to the unavoidable way we move in time towards the future. What is more, since you can get no further once you have reached the singularity, this point must mark the end of time itself! This is where black hole singularities differ from the Big Bang which is a singularity that marks the *beginning* of time. They are more like the Big Crunch singularity

---

[7] From the annoying regular footnotes such as this.

(the one that marks the end of space and time if there were enough matter in the Universe to cause it to collapse in on itself).

The time it takes to reach the singularity from the horizon, as measured by someone falling in, is proportional to the mass of the black hole. Thus for a hole with ten times the mass of the Sun it would take just one ten-thousandth of a second to hit the singularity, whereas for a supermassive black hole it could take several minutes.

A question that is often asked is whether an astronaut falling through the event horizon of a black hole notices anything different (assuming it is a big enough hole for the astronaut to survive the tidal forces). The answer is no. The only way you could find out whether you had crossed the horizon (notice how that astronaut has now become you? don't take it personally, I don't even know you and would not wish such an end on anyone) would be if you tried to halt your fall and climb back out again by firing your rocket engines to push yourself back up away from the centre of the hole. According to the Russian astrophysicist and leading black hole expert, Igor Novikov, just another of the weird aspects of black hole physics and a consequence of the way time is warped is that by trying to do this (firing your rockets to escape from the hole) you will reach the singularity even quicker than if you had left your engines off!

This is certainly very counter-intuitive but he explains it in the following way. Remember that without the rocket engines firing you are in free fall and not feeling any gravitational force (apart from the tidal forces of course). By pointing your rocket away from the singularity and firing the engines you will feel a force of acceleration upwards and, due to the principle of equivalence, this is like feeling the effects of a gravitational field. However, because of the way space and time are mixed up inside a black hole you continue to fall *at the same rate* as before. It is just that now your time will slow down. This is known as gravitational time dilation and I will discuss it in Chapter 6. It means that a fall from the horizon to the singularity that would have taken you, say, ten seconds, might now seem like just five seconds. Weird!

While writing this chapter, I mentioned to my wife, Julie, that I had reached the part where I describe what it is like inside a black hole. "Very dark, I expect" was her deadpan and profound reply. In fact it is not completely dark since the light from the outside Universe still gets in. The difference is that the light becomes bent and focused into a small bright patch. It would be like a view of the receding light from the entrance to a dark tunnel as you venture deeper inside the hole.

Let us now consider what a distant observer sees when an object falls into a black hole that, for simplicity, is assumed not to be spinning. Imagine now that you are in your space ship, hovering at a safe distance outside the event horizon. You witness a colleague falling in towards the horizon. Rather than seeing him falling faster and faster until he suddenly disappears through the horizon, the rate of his fall seems to slow down more and more as he approaches the horizon until he finally stops, frozen, just outside it. This apparent slowing down of a falling object is due to the way gravity affects the rate of flow of time. In fact time literally *slows down* in gravitational fields and this is most noticeable in the strong field outside a black hole.

If the astronaut has calculated that he will pass through the event horizon at twelve o'clock precisely according to both of your previously synchronized watches, then you can, via a powerful telescope, observe the time shown on his watch as he falls. You will see the hands on his watch slow down as he approaches the horizon until they finally stop at twelve o'clock exactly. In fact, at the horizon time stands still. Sometimes it is (wrongly) suggested that you would see him frozen outside the horizon forever. In fact, his image will very quickly fade away and he will disappear. This is not because you have 'seen' him fall through the horizon, but rather because the light reaching you from him has been redshifted to such long wavelengths that it quickly goes beyond the visible spectrum. This redshift is not quite the same as that due to receding distant galaxies whose light is Doppler shifted. Now there is an additional affect due to the slowing down of time near the horizon that makes the light appear to you to have a lower frequency and thus a longer, redshifted, wavelength. The falling astronaut, however, has a different concept of the rate at which time is flowing and calculates that he falls towards the hole faster and faster.

To see a black hole

You may be thinking by now that all this talk of time slowing down, travelling at the speed of light, being stretched like spaghetti then crushed to zero size and infinite density is purely the stuff of science fiction. After all, no one has ever come face-to-face with a real black hole and all these conclusions have been reached by studying their properties theoretically.

Until the 1960s, most astronomers found it hard to believe, despite the theoretical predictions, that there could really be black holes out there. But with the advances in radio and x-ray astronomy and a number of exciting discoveries during the 1960s such as the cosmic background radiation (which confirmed the Big Bang theory), quasars and pulsars, suddenly black holes no longer seemed so outrageous. Coupled with this, many of the important theoretical advances in black hole physics were made during the 1960s and '70s and by the '80s I would guess that astronomers would have been something like 90% sure that black holes existed.

You may consider a 90% confidence level not enough so, thankfully for black hole fans like me, the 1990s have seen a further accumulation of evidence and we are no longer in any real doubt. I would put the current confidence level at 99%. What is this evidence then? After all, if by definition a black hole is black, how can it be picked out against the blackness of space? Even if one happened to have a luminous nebula as its backdrop, you must remember that black holes are so small on an astronomical scale that they would be far too tiny to be seen even by the most powerful telescopes.

The secret to their possible detection (which, amazingly, was pointed out two hundred years ago by John Michell) lies in the way they influence visible matter nearby. Recall that binary stars orbit round each other or, more correctly, around their combined centre of gravity (an imaginary point in space which is the mid-point of their masses). If they have the same mass then they will have the same orbital radius since their centre of gravity will be half way between them, but if one star is much heavier than the other then it will only wobble slightly while the lighter one does most of the 'running' around it. This is because the centre of gravity is now much closer to the larger star.

If one of the stars is large enough to collapse into a black hole then, even though it is now invisible, its gravitational effect on its partner will be the same. Note that they would not be so close together that the other star could get swallowed up by the black hole since the stars would have been attracted together long before if they were that close (but they may still be close enough for the black hole to suck off some of the gas from the surface of its partner).

We should in theory be able to observe the 'wobble' in the motion of the remaining visible star and hence deduce how much mass would be required to cause an object as big as a star to move about like this. After all, single stars do not wobble about for no reason and such motion must be the result of a tremendous nearby concentration of mass. It has been discovered recently that a tiny wobble in a star's position might be due to planets in orbit around it that cannot be seen directly. However, from the mass of the star and the amount of wobble we can deduce how massive the invisible partner is. If it is more than, say, ten solar masses (to be on the safe side) then it would have to be a black hole.

The way the wobble is detected is not, as you might expect, by measuring the sideways motion (the to and fro motion of the star at right angles to our line of sight) but from the change in the wavelength of light that leaves the star when its orbit takes it towards us and when it is receding from us. This is just a Doppler effect again. The wavelength of the light gets compressed (towards the blue end of the spectrum) when it is moving towards us and stretched (to the longer wavelength red end) when it is moving away. It does not matter in what direction the binary system as a whole is moving since it is a *change* in the observed wavelength that we require. From the rate at which this change in wavelength occurs, plus some other pieces of information, we can deduce the period of the orbit and hence the mass of the invisible partner.

This all sounds very clever in theory, but in practice it turned out not to be quite so straightforward. There are other reasons why we would only see one of the stars in a binary system. The simplest explanation is if the other star was just too small and dim and is thus outshone by its larger, brighter, companion. It may be that the invisible partner is a white dwarf or even a neutron star. What is needed is proof that it has a mass that is well above the critical value for a black hole (say five or ten times the mass of the Sun). However, even this is not *proof* that it is a black hole. Igor Novikov puts it this way: "'invisibility' is a poor proof for the existence of something", and he quotes the old joke about the title of the research thesis: 'The absence of telegraph poles and wire in archaeological excavation sites as a proof of the development of radio communications in ancient civilizations'.

What finally clinches it in support of black holes among stellar binary systems is something I have already alluded to. If the two companions (the still shining star and the invisible black hole candidate) are close enough together then the black hole's incredible gravitational field will slowly suck gas off the outer envelope of the star. This gas will spiral in towards the hole's event horizon heating up all the time as it speeds up and forms what is known as an accretion disc surrounding the hole. The matter in this disc is so hot that, before it falls in, it will give off an unmistakable signal of powerful x-ray emissions, which are just bursts of high energy electromagnetic radiation. They will be subtly different from the x-ray emissions produced by some pulsars (the spinning neutron stars) due to their rapid rotation, since in the case of a black hole's accretion disc the timing of the bursts is random. X-ray pulsars give off their signal at regular intervals as they spin, a bit like a searchlight.

Do such x-ray binary systems exist? The answer is yes. The most famous example is one that Stephen Hawking finally, and somewhat reluctantly (due to a bet he had with Kip Thorne), admitted must contain a black hole. It is called Cygnus X-1 and is about six thousand light years from Earth, but within our own Galaxy. The visible companion is a giant star about thirty times the mass of the Sun (deduced by studying the light it gives off). By studying the way it wobbles (from its periodic Doppler shift) the mass of its invisible partner has been put at about ten solar masses[8]. The last piece in the jigsaw comes from the period of the x-ray emissions given off with a frequency of several hundred times per second from the accretion disc and tell us how fast the gas is orbiting the hole. Since nothing can go faster than light, this period gives us a maximum size for the orbit and suggests that the hole must be much smaller than the Earth, and to squeeze ten solar masses into such a small volume means it would have to be a black hole. The laws of physics state that it can be nothing else.

There are several other candidates for black holes within x-ray binary systems in our Galaxy and in nearby ones. It is estimated that our Galaxy alone probably contains millions of black holes!

All I have discussed so far are the common-or-garden black holes that are formed when massive stars collapse under their own weight. There is another type of black hole that is, in a way, even more impressive. Another of the major discoveries in astronomy made in the 1960s was that of *quasars* (which stands for quasi-stellar radio sources). By quasi-stellar it is meant that they were thought to be similar to stars in that they appeared as pointlike objects rather than extended blobs of light like galaxies or nebulae. They also gave off strong radiation in the radio frequency band—and not because they contained their own radio stations as I used to think as a child when I first read about them. Nowadays, only a small fraction of all quasars discovered are radio sources, but the name has stuck. What is more important is that it turns out that quasars are nothing like as small as stars. They are also the most distant objects in the Visible Universe and some are over ten billion light years away (which means that light left them when the Universe was very young). For objects so far away to shine with such brightness means they must be incredibly energetic. In fact quasars are now thought to be young 'active' galaxies with most of their energy (about a thousand times more than the energy output from all the stars in our Galaxy) coming from a tiny central core. This core contains what is known as a supermassive black hole. Typically such black holes would have masses millions of times greater than the Sun.

[8] We are not sure of this estimate and there is still a tiny chance that it might be a massive neutron star if it is at the lower limit of mass it could have.

Since the discovery that quasars have lurking inside them gigantic black holes it has been discovered that many large galaxies probably go through a quasar phase before they settle down. Even if they didn't, they could still well contain a supermassive black hole at their centre. These would have formed from the accumulation of vast amounts of stellar gas in the dense centres of galaxies. Andromeda's black hole is roughly 30 million solar masses with a radius extending out to the size of our solar system. Our own Galaxy's black hole is smaller with an estimated size of just a few million solar masses. Once a supermassive black hole has formed it will slowly feed on the surrounding stars and grow bigger.

Not so black after all

Twenty five years ago Stephen Hawking discovered that black holes are not completely black after all, as viewed from outside. He realized that a quantum process known as pair creation can cause a black hole to leak its energy very slowly out into space. As it does this, it will gradually shrink in size until it finally explodes and nothing is left. It is therefore not true to say that what falls into a black hole never comes out. It will, eventually, be radiated out as tiny particles over a period that, for all intents and purposes, can be considered to take forever!

This process is called Hawking radiation and requires a little explanation.

In the subatomic world, two particles, such as an electron and its antimatter particle[9] (called a positron) can spontaneously pop into existence out of complete nothingness. Quantum mechanics says this is allowed as long as the two particles recombine very

---

[9] A common misconception is that antiparticles have negative mass and so this cancels the positive mass of the particle. This is not the case. Antimatter has the same kind of mass, and is affected by gravity in the same way, as normal matter. The 'anti' bit is to do with these particles having opposite electric charge (plus a few other differences). So since electrons are negatively charged, positrons are positive. Otherwise they are identical in every way apart from the fact that our Universe contains mostly electrons.

quickly in a process called pair annihilation, to give back the energy they must have needed to form in the first place. I know this will not make sense. In fact, it should not make sense, but many strange things can and do happen in the world of quantum mechanics. You may well ask where the energy to create the particles came from in the first place. The answer is that the energy can be 'borrowed from nowhere' and must be given back to 'nowhere' very quickly since nature does not enjoy being in debt for long. Hawking realized that when a pair of particles is created very close to the event horizon of a black hole it could be that either the particle or its antiparticle might fall into the hole while the other is able to escape. Since it can no longer annihilate with its friend, it is allowed, like Pinocchio, to become a real live electron or positron. The energy it has kept and does not have to pay back has come from the black hole itself.

Hawking radiation must go on all the time just outside the horizon of black holes, but only in a very small number of cases will one of the particles escape. Most of the time both will fall into the black hole. The reason one of the particles can ever get away is due to tidal forces: the particle that is slightly closer to the horizon is pulled much more strongly towards it.

For normal-sized black holes this process can be completely ignored since far more particles will be sucked in from the space surrounding the horizon than are ever lost through Hawking radiation. The effect only becomes significant enough to be interesting when a black hole has shrunk down to microscopic size (after a time much longer than the life of the Universe) when the region outside its horizon heats up and the radiation gets very intense. This shrinking of a black hole is due to it obeying Einstein's famous equation $E = mc^2$ which states that mass (m) and energy (E) are interchangeable and one can be converted into the other. This process is happening twice here. First, the escaping particle has been endowed with mass (given substance) converted from pure energy which has been extracted from the black hole. Then, the hole itself, having lost this energy by being responsible for the creation of a particle, must produce that energy by giving up a minute fraction of its own mass. Viewed from afar we would say that the black hole has spat out a particle. Its event horizon appears hot due to its radiation of these particles. But it is only around very tiny black holes that this process is non-negligible.

It has been suggested that such mini black holes may actually exist. They would have been created just after the Big Bang and some might still be around today. As an example, if Mount Everest could be squeezed down to the size of an atom it would be converted into such a mini black hole. It would then give off Hawking radiation at a very high rate. Even so, it would not completely evaporate for billions of years. Now it just so happens that the Universe is about 15 billion years old, so any mini black holes created in the early Universe with an initial mass equal to that of Mount Everest will just about have completely evaporated away by now. Since they radiate at an increasing rate as they shrink, they should end in a tremendous final explosion of high energy radiation. Believe it or not, astronomers are on the look-out for such telltale bursts of radiation.

It may be that the energy of black holes can be extracted artificially too. In this case it is the rotational energy that is milked. The English mathematical physicist, and long-time collaborator of Hawking, Roger Penrose, had proposed, even before Hawking's evaporating idea, that if an object enters the ergosphere of a rotating black hole and then splits into two parts with one falling into the hole, then the other half can be thrown clear with more energy than it came in with. The energy it acquires has come from the hole and will slow its rotation down slightly.

It may be that an advanced civilization that comes across a black hole could utilize this method to turn it into an energy source.

White holes

There are certain solutions of the equations of general relativity which allow for the existence not only of black holes but of objects called white holes too. But it turns out that this would be possible only if the Universe had certain properties, called initial conditions, at the time of the Big Bang. These highly speculative objects are the opposites of black holes which, rather than sucking matter in, would spew matter and energy out into the Universe. Their singularities would be different too and would mark the beginning rather than the end of time. Unlike black holes though there has been no conclusive evidence that white holes actually exist. One of the problems is that the matter leaving them might fall back in and the white hole would quickly be converted into a black hole.

It is now believed that large white holes do not exist in the Universe, but the rules of quantum mechanics regarding particles and antiparticles suggest that if mini black holes exist on the subatomic scale, then so should their antimatter partners: mini white holes. Hawking has suggested that, just as particle/antiparticle pairs can pop into existence for a fleeting moment, so it should be possible for a black hole/white hole pair to suddenly appear from nowhere. But don't worry, they would be too tiny to have any effect on us.

TIME

5

TIMES ARE CHANGING

"Time is nature's way of keeping everything from happening at once."
John A Wheeler

"Time is just one damn thing after another."

Unknown

Foreigner in London: "Please, what is time?"
Man on street: "That's a philosophical question. Why ask me?"
J G Whitrow

What is time?

Let's get one thing straight. Whatever you have read or heard, nobody understands what time really is. There has been so much written about the nature of time, particularly over the past few years, that it would be difficult in this book to contribute much that is original or that has not been discussed elsewhere. But that is not my intention. I do not feel as though I need to have trawled through the many excellent books (and some less so) that have dealt with the subject of time—although I have read a fair few over the years—and then to try and come up with some new 'angle' or clever argument not used before: my theory of time. Of course a lot of what has been written about time is utter nonsense, but there is much that, despite sounding like nonsense the first time

111

you come across it, actually makes some sense provided you are prepared to give it some thought.

I mentioned at the beginning of the book that the subject of time had fascinated me as a child, and still does. I am not alone. In fact I am probably in the majority. The sad fact is that I am no closer today than I was at the age of ten to understanding what time really means. I understand how many of the laws of physics contain time in a fundamental way, I have heard many of the philosophical arguments about the flow of time, the direction of time, whether time is really 'out there' or is just an illusion: a construct of human imagination. But whether I am any more enlightened is debatable.

One thing is for sure though. Like so much else we have seen thus far, Einstein's theories of relativity at the beginning of this century overthrew the old and cherished notions. I will discuss the relativity of time in the next chapter. For now I will lead you through some of the ideas in physics and philosophy about the nature of time, most of which were around long before Einstein.

Who invented time?

Humans have long been aware of the cyclic nature of time in the regular way that night follows day and the passing of the seasons. We are also aware of the linear nature of time flowing from past to future. Events that are now in our past will remain there never to return but will recede further and further back.

Early in the history of mankind, it became necessary to divide up a day into smaller units of time. Since the motion of the Sun across the sky—long before it was known that this was due to the Earth's rotation—took (roughly) a fixed amount of time it is not surprising that one of the earliest timekeepers was the sundial, invented over five thousand years ago in ancient Egypt.

The big change to mechanical clocks came in the sixteenth century when Galileo discovered that a pendulum of a given length will always take the same time to complete one full swing. But it was not until the mid-seventeenth century that the first pendulum clock was built. This accuracy allowed for much more precise timekeeping than before, and hours were divided into minutes and minutes into seconds. Nowadays, pendulum and clockwork timekeepers are slowly being replaced by ever more reliable ones. A digital watch contains a tiny quartz crystal which vibrates thousands of times a second when electricity is passed through it. These vibrations are so regular you could set your watch by them. (Sorry!) Can you imagine how difficult life would be for us today with its appointments, schedules and deadlines if the smallest unit of time we had was the hour?

Today the most accurate timekeepers in the world are atomic clocks which can measure intervals of time with extraordinary precision. They rely on the fact that certain atoms, when pumped with energy, emit light at a precise frequency that is unique to that type of atom. The most famous of these are caesium clocks which now set the world standard for time.

While the 'second' is the standard unit of time, it is clearly a human invention. If there is intelligent life elsewhere in the Universe they would measure time using their own 'currency' which could well derive from the time it takes for their home planet to complete one revolution or one orbit round their sun. Until recently, our 'second' was defined as one sixtieth of a sixtieth of a twenty-fourth of the time it takes the Earth to complete one revolution around its axis (i.e. a day).

This *was* how one second was defined, but not any more. These days, we are so obsessed with time that this definition is no longer adequate. You see there is a problem. It turns out that the Earth is slowing down. Not enough so as you'd notice, just a second every few years, but this is enough to mean that, in our high tech world, we need another way of measuring time. So, since all atoms of caesium always radiate light that has a frequency of 9,192,631,770 cycles per second, scientists decided that they would turn the statement around and say that one second is *defined* to be the interval of time required for light from caesium atoms to oscillate 9,192,631,770 times. This is called co-ordinated universal time. The length of one day according to co-ordinated universal time is therefore $24 \times 60 \times 60 \times 9,192,631,770$ vibrations of a caesium atom. This has meant that every few years we must add in a leap second to take into account the slowing down of Earth's spin so that the new definition of time does not drift away from the old one.

What about time as a concept in itself, rather than how we humans measure it? Until Isaac Newton completed his work on the laws of motion, time was considered to be the domain of philosophy rather than science. However, Newton described mathematically how objects move under the influence of forces, and since all movement and change requires the notion of time for it to make sense, he used what is known as a realist view of time. This 'common sense' view is still with us today, despite the fact that we know it is wrong as we shall see in the next chapter.

Newtonian time is absolute and relentless. He described it as a medium which exists entirely on its own outside space and independent of all processes that occur within space. In this view, time is said to flow at a constant rate as though there were an imaginary cosmic clock that marks off the seconds, hours and years regardless of our, often, subjective feelings about its passage. According to Newton, time is absolute, true and mathematical. We have no influence over its rate of flow and cannot make it speed up or slow down. We also know how unreliable we sometimes are at judging intervals of time. Imagine you were to drop off to sleep on a train journey that normally takes one hour and you wake up feeling that only about ten minutes have elapsed. When you check your watch you see that it is a whole hour later and this is confirmed when you look out the window to see that you are close to your destination. Of course it could be that your watch malfunctioned and that the train speeded up considerably while you slept for what really was only ten minutes, but this is highly unlikely since we know how unreliable human subjective time keeping can be. We all have this gut feeling that Newtonian time is really 'out there' and flows at the same rate everywhere in the Universe.

The world's major religions all have something to say about the nature of time. The monotheistic religions believe in an omnipotent God who created the Universe and who exists outside our space and time. He is omniscient in the sense that He knows

114

not only the past but the future, and He is omnipresent in being in all places at all times. An eternal God who therefore exists outside our Universe does not conflict with the notion in modern physics of the Universe (which includes space *and* time) coming into being at the Big Bang.

What has been a major topic of debate among scientists, philosophers and theologians, however, is the part God plays in Newton's deterministic clockwork universe. According to the mechanistic view that Newton's laws of motion give us of the Universe, it is possible, in principle at least, to know the position and velocity of every particle in the Universe. Given that each particle will follow a well-defined trajectory and be under the influence of forces that, again in principle, can be well defined, it is possible to work out their positions at any future time and hence to know the state of the Universe in the future. The future is therefore mapped out and preordained.

Such a reductionist view of the world might seem to leave no room for human free will. Since we too are made up of atoms we are subject to the same laws of physics as any other object; then presumably what we consider to be free will is no more than mechanical processes in the brain obeying Newton's laws like everything else.

In practice of course we are not even able to calculate the future positions of a few balls on a pool table after they are scattered by the cue ball, let alone the future positions of all the particles in the Universe. But, according to this 'deterministic' view it should at least be possible *in principle* to do so, provided we had a powerful enough computer. Such a computer would have to run a program of such stupendous complexity that it would contain many more unknown variables than there are particles in the Universe. This is because each particle needs (at least) six numbers to define its state at any given time: the three that tell us where it is in 3D space and three more to tell us how fast it is moving and in which direction.

To a good approximation, we would not need all this information since an atom in a distant galaxy is not going to affect things on Earth, but even if we restrict ourselves to the atoms on Earth we are still dealing with a pretty impressive number. After

115

all, there are more atoms in a single glass of water than there are glasses of water in all the oceans of the world.

Nevertheless, I emphasize that, as long as the number of particles we are dealing with is not infinite, then we can consider an imaginary computer that could calculate the future position of all the particles in the Universe if it knew what they were doing now. And knowing the future implies knowing what all bodies are going to do next. Such knowledge should extend to humans too since we are all only made up of atoms.

Today physicists no longer adhere to this idea of a deterministic universe. That way of thinking was overthrown when the theory of quantum mechanics was developed in the mid-1920s showing that, at its most fundamental level, nature is random and unpredictable[1]. Despite this, many physicists believe that the future *is* already out there, not because of the Newtonian picture of a clockwork universe, but because it follows on from the way the theory of relativity unifies time with space. This idea that the future already exists goes beyond the Newtonian view, which only claims that the future *can* be predicted.

As for the nature of time, not everyone was happy with Newton's realist view of an external absolute time, even before the twentieth century's two scientific revolutions in modern physics of relativity theory and quantum mechanics. Scientists, philosophers and theologians have long debated several issues which I will discuss briefly here. They concern the three concepts of the origin of time, the flow of time and the direction of time.

The first moment

I will first deal briefly with the question of the origin of time. Most present day religions teach of a moment of creation when

---

[1] I am aware that the use of the word 'unpredictable' is misleading here. Quantum mechanics is only unpredictable in the same way that the toss of a coin is unpredictable. If, however, you toss a coin a hundred times then roughly half of the outcomes will be heads and the other half tails. So there is a definite statistical rule which applies for a large number of trials. Quantum mechanics can be thought of in this way.

the Universe came into being. They may differ from each other in the 'how', 'why' and 'when', but the basic idea is the same. As we saw in Chapter 3, most physicists (some of whom are themselves devoutly religious) now also believe that the Universe began at a definite moment, about 15 billion years ago. But can we say that the Big Bang 'happened' at some definite moment in time?

The problem is that time itself is thought to have started at the Big Bang and is part of the fabric of the Universe. The Big Bang cannot even be considered as the 'first event' since that would require it to have happened within time. This idea is not unique to science and many religions have a Creator who exists outside time, leaving Him free to create time itself.

Physicists are now trying to understand *why* the Big Bang happened in the first place. What caused it? Unfortunately, cause and effect are notions that require time, and since the Big Bang marked the beginning of time we cannot say that something 'prior' to it caused it. It may have just 'happened'.

And as if this is not enough, remember that in order to understand the world of the very small we need to apply the ideas and concepts that arise from the theory of quantum mechanics, and you don't get much smaller than singularities. The Big Bang singularity must therefore be treated as a quantum 'event'. Physicists have yet to sort out many of the details but nevertheless argue that, in the quantum world, things get fuzzy and indefinite, even the ordering of events. Strangely enough (or conveniently enough depending on your viewpoint) quantum mechanics allows things to happen without a cause, including the Big Bang itself.

One explanation which physicists are fond of using to describe how the Universe came into existence is that the rules of quantum mechanics would have allowed the Big Bang to happen on the understanding that the Universe should quickly 'pop back out' of existence again. For reasons we do not fully understand yet, what may have happened next was that the Universe quickly underwent a brief period of extremely rapid expansion after which it became a permanent fixture, still expanding but at its current, more sedate, rate.

So if it wasn't the Big Bang, what was the first event in the now created Universe? Physicist and author Paul Davies, who has delved as much as anyone into the nature of time, explains that there cannot even have been a first event. He likens it to asking what the first number is after zero. We must consider all numbers and not just whole ones (the integers) otherwise the first number after 'zero' would be 'one'. It does not matter how small a number we choose, we can always halve it to get a smaller one. In the same way, there was no first event after the Big Bang. However early the event, there will always have been an earlier time closer to the Big Bang to consider.

However, as soon as quantum mechanics is brought into the debate, we find that there is indeed an 'earliest time' after the Big Bang. At the tiniest length and time scales, everything gets grainy and fuzzy, including time itself. Just as the concept of the ordering of events no longer applies at these extremes, neither does the idea of continuous time. At this scale, an interval known as the Planck time can be considered as the shortest possible meaningful slice of time. Of course we are unaware of such a departure from the smooth flow of time because the Planck scale is so tiny. In fact, there are unbelievably more units of Planck time in one second than there have been seconds since the Big Bang. Anyway, the point is that if you go back in time to one unit of Planck time after the Big Bang, it makes no sense to ask what happened before it.

Does time flow?

Many philosophers have argued that time itself is an illusion. Consider this: time consists of past, present and future. Even though we have records of the past and memories of certain events that have taken place, it can no longer be considered to exist. The future on the other hand has yet to unfold and therefore does not exist either. This leaves the present which is defined as the dividing line between past and future. Surely the 'here and now' exists. But although we 'feel' that this line is steadily sweeping through time gobbling up the future and converting it into past, it is nevertheless just a line and as such does not have any thickness. The present, therefore, is of zero duration and cannot have a real existence either. And if all three components of time do not exist then time itself is an illusion!

You may, as I do, take such clever philosophical arguments with a pinch of salt. What is much harder to justify, however, is the notion that time 'flows'; that time goes by. It is hard to deny the feeling that this is what happens, but having a 'gut' feeling about something, however strong that feeling is, is not enough in science. In our everyday language we say that 'time passes', 'the time will arrive', 'the moment has gone' and so on. But if you think about it, all motion and change must, by definition, be judged against time. This is how we define change. When we wish to describe the rate of a certain process we either count the number of events in a unit of time, such as the number of heart beats per minute, or the amount of change in a unit of time, such as how much weight a baby has put on in one month. But it becomes nonsensical to try and measure the rate at which time itself changes since we cannot compare it with itself. People often jokingly state that time flows at a rate of one second per second. This is clearly a meaningless statement since we are using time to measure itself. To clarify this, how would we know if time were to suddenly speed up? Since we exist within time and measure the duration of intervals of time using clocks which, like our internal biological clocks, must presumably speed up also, we would never be aware of it. The only way to talk about the flow of (our) time is to judge it against some external, more fundamental, time.

But if an external time against which we could measure the rate of flow of our own time did exist then we would only be pushing the problem further back rather than resolving it. Surely if time by its nature flows, then why should this external time not flow also? In which case we are back to the problem of needing a further, even more fundamental, time scale against which to measure the rate of flow of external time, and so on in a never-ending hierarchy.

Just because we are unable to talk about a *rate* of flow of time does not mean that time does not flow at all. Or maybe time is standing still while we (our consciousness) are moving along it (we are moving towards the future rather than the future coming

119

towards us). When you look out of the window of a moving train and observe fields rushing by you 'know' that they are standing still and that it is the train that is moving. Likewise, we have the strong subjective impression that the present moment (what we call now) and an event in our future (say next Christmas) move closer together. The time interval separating the two moments shrinks. Whether we say that next Christmas is moving closer to us or that we are moving closer to next Christmas amounts to the same thing: we feel that something is changing. So how come most physicists argue that even this idea is not valid?

Strange as this may sound, the laws of physics say nothing about the flow of time. They tell us how things like atoms, pulleys, levers, clocks, rockets and stars behave when subjected to different forces at certain instants in time, and if given the status of a system at a particular moment the laws of physics provide us with the rules for computing its likely state at some future time. Nowhere, however, do they contain a hint of flowing time. The notion that time *passes*, or moves in some way, is completely missing in physics. We find that, like space, time simply exists; it just *is*. Clearly, say most physicists, the feeling we have that time flows is just that: a feeling, however real it may seem to us.

What science is unable to provide at the moment is an explanation for where this strong sense we have of passing time and a changing present moment comes from. Some physicists and philosophers are convinced that there is something missing in the laws of physics. I will not go as far as to say that I subscribe to this view, but I do believe we will only make progress when we have a better understanding of how our own consciousness works, and hence *why* we feel the passage of time.

I should mention that no less an authority than Einstein himself held the view that the flow of time is illusory and even expressed it when trying to console the bereaved widow of a close friend of his, stating that she should take comfort in the knowledge that the present moment is no more special than any other in the past or the future; all times exist together.

Something called entropy

Even if time does not flow, we can still assign to it a direction, called an arrow of time. This is an abstract concept which simply means that we can define an ordering of events. An arrow of time points from the past towards the future, from earlier events to later ones. It is a direction in time in which things happen. It is important here to make the distinction between a flow of time and a direction of time. Imagine looking at individual frames on a reel of a movie. We can easily define an arrow of time pointing in a particular direction along the reel based on which frames were earlier and which were later. We do this despite the fact we are looking at still shots of events and there is no movement in the frames. Each one is a snapshot frozen in time.

Even when it comes to the direction of time we must be careful. We must not confuse the real direction of time (if any such thing exists) with our subjective feeling for the direction of time. Let me first define what may appear to be an obvious arrow of time, known as the *psychological arrow*, which is the direction that we *perceive* time to point in; the fact that we remember events in our past and look ahead to events that have yet to happen in our future. If your psychological arrow of time were to suddenly flip over it would appear as though everything around was running in reverse. Everybody else's future would be in your past and vice versa. This is clearly so ridiculous that I will not waste any more time discussing it and you can stop trying to make sense of it. Is there indeed a problem with the arrow of time at all? Surely the fact that we see the past happening before the future is because the past *does* happen before the future!

The reason why I am being cautious here is that the equations of physics do not even provide a direction in time. Time could flow backwards and the laws of physics would stay the same. You might argue that this is just tough luck for physicists. If the direction in which time should point is missing from the equations of physics then they cannot be telling us the whole story. Just because they cannot discern a direction for time from the mathematics does not mean that there isn't one in the *real* world.

121

But the problem is more serious than this. Even in the real world, at the level of atoms, almost all processes are reversible in time. If, in a subatomic process, two particles, a and b, converge and collide they will often bounce off each other and separate again. If you were to watch a film of such a process and then watched it running in reverse, you would not be able to decide which way round the process happened. The time-reverse process still obeys the laws of physics. I should point out that this would have to be a thought experiment. We could not really do it since no microscope on Earth is powerful enough to resolve detail down at the subatomic level.

It often happens that instead of the same two particles bouncing off each other, two new ones, say c and d, are produced and fly apart. Again, you would not be able to decide on the true order of events if you watched a film of this process because the laws of physics state that the reverse process is also possible. Particles c and d could have collided to produce particles a and b. You therefore cannot assign an arrow of time that would state which way round the process occurred.

This is in sharp contrast with events that happen around us in everyday life where we have no trouble deciding which direction time is pointing. For instance, you never see smoke above a chimney converging on it and getting neatly sucked down inside it. Similarly, you cannot 'unstir' the sugar from a cup of coffee once it has been dissolved, and you never see a pile of ash in the fireplace 'unburn' to become a log of wood again. What is it that distinguishes these events from the subatomic ones? How is it that most of the phenomena we see around us could never happen backwards? Surely everything is ultimately made up of atoms and at that level everything is reversible. So at what stage in going from atoms to chimney smoke, cups of coffee and logs of wood does a process become irreversible?

On closer examination we see that it is not that the processes I have described above can never run in reverse, but rather that they are extremely unlikely to do so. It is entirely within the laws of physics for dissolved sugar to 'undissolve' through stirring and reconstitute itself into a sugar cube again. But if we ever saw this

122

happening we would suspect some kind of conjuring trick. And rightly so, for the chances of it happening are so tiny they can be ignored.

Let us consider a simpler example using a pack of cards. It is simpler because we are dealing with a much smaller number of components (fifty two cards) than the number of molecules of sugar or smoke or wood in the above examples. Begin with a pack of cards which has been ordered such that the four suits are separated and the cards in each suit are arranged in ascending order (two, three, four, ..., jack, queen, king, ace). By shuffling the cards a little the order will be ruined. Now we can ask what happens to the order of the cards upon *further* shuffling? The answer is obvious: it is overwhelmingly more likely that the cards become even more mixed up than it is for them to return to their original ordered arrangement. This is the same irreversibility as in the case of a partially dissolved sugar cube which on further stirring always carries on dissolving.

To give you an idea of the probabilities involved, if you were to take a completely shuffled pack of cards then the chances of getting the ordered arrangement you started with through further shuffling is about as likely as it would be for you to win Britain's National Lottery jackpot not once or twice but on nine consecutive draws!

It is all down to an important law in physics called the second law of thermodynamics. The subject of thermodynamics involves the study of heat and its relation with other forms of energy. The astronomer Arthur Eddington went so far as to claim that the second law held the supreme position among all the laws of nature. There are three other laws of thermodynamics which are to do with how heat and energy can be transformed into each other, but none is as important as the second law. It has always amused me that one of the most important laws in the whole of physics cannot even make it to the number one spot on the list of thermodynamics laws.

The second law of thermodynamics states that things wear out, cool down, unwind, get old and decay. It explains why the sugar dissolves in the coffee but never undissolves. It also states that an ice cube in a glass of water will melt because heat is always transferred from the warmer water to the colder ice cube and never in reverse. To understand the second law a little better I must introduce you to a quantity called *entropy*. The second law is a statement of increasing entropy. In an isolated system, entropy will either stay the same or increase, but can never decrease.

Entropy is a quantity which is a little difficult to define precisely so I will do so in two ways:

1. Entropy is a measure of untidiness in a system; how mixed up things are. The ordered pack of cards described earlier is said to have low entropy. By shuffling the pack we are ruining their initial order, and increasing the entropy. When the cards are completely mixed up the entropy of the pack is said to be at its highest and further shuffling cannot mix them any more[2].

2. Entropy can also be thought of as a measure of a something's ability to do work (by which I mean the possibility of extracting useful energy from it rather than work in the usual meaning of the word). A fully charged battery has low entropy which increases as the battery is used. A clockwork toy has low entropy when wound up which increases as it unwinds. When it has completely unwound, we can reset its entropy back to a small value by winding it up again. The second law is not being violated here because the system (the clockwork toy) is no longer isolated from its environment (us). The toy's entropy is being decreased but we are 'doing work' to wind it up and our entropy is increasing. Overall, the entropy of toy + us is increasing.

It is a little difficult to provide an example of entropy which encompasses both of the above definitions: that of increasing disorder and that of the ability to do work. However, one such example of the unavoidable increase in entropy is my children's bedrooms. Before they get back from school in the afternoon their rooms are tidy and said to be in a state of low entropy. Once they are home and playing behind closed doors there is an impressively rapid rise in entropy. Lego bricks, cars, dolls, teddy bears, plastic

---

[2] Of course further shuffling will mix the cards up in a different way, but we would not be able to say that they are any more mixed up.

tea sets and an assortment of plastic food all get pulled out of their boxes and strewn randomly across the floor. The only way to get the rooms back to their initial low entropy state is to 'apply external work to the system' (usually in the form of their mother). It would be against the laws of physics (or what is known as the second law of the Al-Khalilis) for the children to enter a high entropy bedroom and, without any external work (such as verbal threats) to decrease its entropy.

Another example of increasing entropy is cigarette smoke in the library canteen at my university (the last refuge for smokers on campus). When a cigarette is lit in the smoking area entropy is said to be low since the smoke is neatly confined to a small volume of the canteen. But thanks to the second law of thermodynamics we are all soon sharing the fumes. The second law of thermodynamics states that you never witness smoke that is evenly distributed around the canteen collect back in the corner again.

We sometimes see examples where it appears as though entropy is decreasing. For instance, a wristwatch is a highly ordered and complex system that is produced from a collection of bits of metal. Surely this is violating the second law. In fact this is just a more complicated version of the example of the clockwork toy. The watchmaker has put a certain amount of effort into making the watch, increasing his own entropy slightly. In addition, smelting the ores and machining the metals that are needed have produced a certain amount of waste heat that more than compensates for the small decrease in entropy due to the creation of the watch.

If it ever seems like entropy is decreasing we always find that in fact the system under consideration is not isolated from its surroundings and that, by zooming out to view a wider picture, the entropy will always be greater than it was before. We can view many processes that happen on Earth, from the evolution of life to the building of highly ordered and complex structures, as reducing the entropy on the surface of our planet. Everything from cars to computers to cabbages has lower entropy than the raw materials it is made up from. Despite this, the second law is not being flagrantly disregarded. What we are missing is the fact that even the whole Earth cannot be considered as isolated from its surroundings. We must not forget that almost all life on Earth, and hence all low entropy structures, is thanks to sunlight. When we consider the combined Earth + Sun system we see that the overall entropy is increasing because the radiation that the Sun pours out into space (only some of which is absorbed by the Earth) means that its entropy is increasing by much more than the corresponding decrease on Earth.

Arrows of time

Where is all this taking us? I began by talking about the direction in which time flows. Remember this is not a true direction in the sense of North or South, or even a direction *in time*; it is a direction *of* time and can only point in one of two (opposite) directions. There are two ways of choosing such an arrow: we can either consider two events and ask which one happened first or, by considering a quantity that is changing, we can choose an arrow of time to be pointing in the direction of increase or decrease of that quantity.

It is often claimed that the reason we 'see' time flowing in the direction that we do is because our brains, like any other physical system, must obey the second law of thermodynamics. Thus the psychological arrow of time must always point in the direction of increasing entropy. This is extremely dubious. To suggest that the entropy in our brains is increasing is wrong. Like any other biological system, our brains utilize energy to maintain their low entropy state. To a good approximation, the entropy in our brains remains constant for most of our lifetime.

The second law of thermodynamics gives us an arrow of time which seems to be more general and less subjective than the psychological arrow of time that you and I seem to have built into our consciousness. We therefore define what is called a *thermodynamic arrow of time*, which always points in the direction of increasing entropy. Since we always see entropy around us increasing, then *by design* the thermodynamic arrow will point in the same direction as the psychological arrow.

What if one day entropy began to decrease everywhere in the Universe? We would say that the thermodynamic arrow has flipped over. What then happens to the psychological arrow? Does it now point in the opposite direction? Do we now see sugar undissolving, packs of cards unshuffling and cigarette smoke collect up from all around a room and focus in on, and disappear into, the tip of a lit cigarette?

The answer, some believe, is no. It is here that they appeal to the notion that our thought processes, which define the psychological arrow, are chemical processes in the brain, and like any other physical system, should be subject to the second law. If for whatever reason, entropy begins to decrease *everywhere*, then that includes our brains (and thought processes) and the psychological arrow would flip over too. I am not so sure because, as I mentioned earlier, I believe that our brains strive against the tide of increasing entropy outside. It is far from clear to me what would happen inside our brains if entropy began to decrease everywhere else.

There are two further arrows of time I should mention which reflect different types of irreversible process in physics. The first is the quantum measurement arrow. As long as a quantum system, such as an atom, is left alone and we do not attempt to measure its properties, it remains fully reversible in the sense that processes that go on inside it could happen forwards or backwards in time. However once we attempt to probe the system (using some experimental apparatus such as a detector to measure the position of an atom say) a definite direction in time is chosen. Certain properties are permanently altered by the act of measurement.

Recent research into the meaning of quantum mechanics suggests that the quantum measurement arrow is very similar in origin to the thermodynamic arrow. Another way of defining increasing entropy is through loss of information. By saving a file on computer you are creating order and decreasing entropy locally. The reverse happens when you erase a file. You are losing information and entropy increases. It is now emerging that the quantum measurement arrow comes about because of a similar loss of information on the subatomic level. In technical

127

jargon it is said that quantum coherence leaks out into the environment surrounding the quantum system when it is probed, thus increasing its entropy. This loss of quantum information is a bit like the way a hot object leaks heat out into its cooler environment.

Finally, I should mention a fourth arrow of time in the light of recent experimental findings. It is called the matter/antimatter arrow. In a rather subtle experiment carried out at the CERN particle accelerator in 1998 it was discovered that it is slightly more likely for antimatter to convert into matter than the other way round. The experiment, known as CP-LEAR (which stands for charge parity experiment in the low energy antiproton ring) is not cut and dried. Rival research groups around the world have yet to be convinced. But if correct, it suggests that if you were to start off with an equal amount of matter and antimatter, in the form of subatomic particles called kaons, then at a later time there should be fewer antimatter kaons than normal matter kaons. This provides us with an arrow of time at the level of these particles, pointing in the direction of diminishing antimatter.

Stephen Hawking gets it wrong

Soon after I started my PhD in 1987 I was in my university library carrying out what is known as a literature search. I was working on a problem in physics which involved a lengthy mathematical calculation describing what happens when two atomic nuclei collide, and I was looking up some references in scientific journals related to my work. Not having been very successful in locating a particular paper and becoming a little bored I decided to look for any recent scientific papers by Stephen Hawking, for no reason other than that I felt his work on cosmology might provide a welcome break from mine. I found a paper of his dating back a couple of years to 1985 in which he discussed how the direction of time might get switched round if the Universe ever began to contract. This sounded promising. I made a photocopy of the article and read it on the train home.

I followed the arguments of the first few pages but very soon became stuck in the mathematics. Nevertheless, that evening I decided that he had to be wrong, but since I could not follow the mathematical details I did not feel I was on safe enough ground. After all, he was a world famous scientist and I had just started as a research student in a different field of physics altogether. Although I did not know it at the time, Hawking had already realized that his conclusions in that paper, which had attracted considerable attention, had been wrong. I wish, nonetheless, to discuss the ideas involved mainly to show just how confusing and illusory time can be if someone of the stature of Stephen Hawking could get things wrong. In fact, it has been fascinating for me to see how so many other renowned scientists and world experts can still hold completely opposite views about something as fundamental as this[3]. It is all down to the confusion many have with the concept of entropy. I shall first briefly describe why Hawking reached his controversial conclusion.

The second law of thermodynamics should apply without discrimination everywhere in the Universe stating that the entropy of any isolated system cannot decrease. So why shouldn't it apply to the whole Universe? After all, the Universe in its entirety is by definition an isolated system since there is nothing outside it. In fact, the entropy of the Universe is indeed increasing and implies that it must have been more ordered in the past. In fact, it must have had a minimum entropy at the Big Bang and has been unwinding, or running down, ever since.

Of course you may consider it rather ambitious, if not arrogant, of us to talk about the entropy of the whole Universe, but since we are trying to figure out its size, shape and age, why not its entropy too? To begin with, I will consider a simple 'model' universe that has little to do with reality but will help us understand what role the second law might play in the evolution

---

[3] Stephen Hawking's original conclusion and his subsequent admission of his error are well documented. But there are other, equally prominent, physicists who have not made any public recantation after their theories had been disproved due to a lack of Hawking's honesty and integrity. Hawking himself states that "there ought to be a journal of recantations in which scientists could admit their mistakes. But it might not have many contributors".

of the Universe. Imagine a sealed box in which all the molecules of air are concentrated together in one corner. One way to achieve this is if all the air is first confined inside a bottle in the corner which can then be opened remotely. The entropy of the box in this initial state is low since the contents are in a highly ordered state with all the air tidily contained within the bottle.

As time goes by the air will escape from the bottle and spread out to fill the whole box causing its entropy to increase. When the air molecules are evenly distributed throughout the box, entropy will be at a maximum and the system is said to be in equilibrium. This is equivalent to the pack of cards being completely shuffled. There is an exceeding tiny probability that at some later time we would find all the molecules back inside the bottle again.

Now imagine the box is much larger (say the size of a galaxy). With so many molecules inside the box their combined mass is enough for gravity to have an effect. It may happen that a group of them could randomly drift closer together than average. Once this happens we would expect gravity to take over and cause them to be attracted towards each other. The more molecules that clump together, the more effective their combined gravitational pull on the surrounding molecules will be. This gravitational clumping will ultimately cause all the air to be clustered together in heaps of different sizes throughout the volume of the box, with empty gaps in between. What has happened to entropy now? We started with the molecules distributed evenly throughout the volume and entropy at a maximum and ended up with what looks like a more highly ordered state, like sweeping Autumn leaves up into neat piles. It looks like gravity has caused the second law to be violated.

Not so. If you think of increasing entropy as a winding down process, then matter that is close enough together to feel the pull of gravity will 'unwind' as it gravitates together. A ball at the top of a hill has low entropy. When it rolls down the hill (due to the action of gravity) its entropy increases. We say that it is losing the ability to perform work. We are taught at school that the ball at the top of the hill has potential energy which turns into kinetic energy as it rolls down. In the same way, a wound up toy (which I described earlier as having low entropy) has potential energy which it loses as it unwinds and its entropy increases.

So gravity increases entropy, but this still doesn't explain how entropy in the box could increase if it was at a maximum already. The answer is that all the time the molecules are evenly separated, gravity will be pulling in all directions and cancelling out, and entropy is at a maximum. If, by chance (and anything is possible), the molecules in a certain region find themselves closer together than average then this represents a temporary departure from the maximum entropy (equilibrium) state. In order for the second law to rectify the situation, these molecules have two choices: they can either drift back apart again to their original equilibrium state, or they can gravitate together to form a clump. Either way entropy increases back to a maximum again. Both scenarios can be seen as a running down of the system, but now we have two alternative pictures of the maximum entropy state.

Now we are ready to tackle the real Universe. Hawking began his argument by stating that the Universe must have had a minimum entropy at the Big Bang and, since it must obey the second law of thermodynamics, has been unwinding ever since, moving towards a state of maximum entropy. He had developed a theory of the Universe which required it to be closed and believed that it contained enough matter to one day halt the expansion and cause it to collapse to a Big Crunch. Recall from Chapter 3 that this is one possible scenario for the fate of the Universe that we cannot rule out. The rest of this chapter will assume that this will indeed be the fate of the Universe (which, as we have discovered, is not likely now).

In Hawking's model, the Big Bang and Big Crunch singularities were identical. After all, in both cases all the matter and energy in the Universe would be crushed to infinite density and zero size. Thus if the Big Bang singularity was in a state of low entropy, then so should the Big Crunch singularity be. It therefore followed that, as the Universe contracted, its entropy would have to decrease again and the second law of thermodynamics would be violated during this phase. Hawking believed that the state of maximum expansion also represented the state of maximum entropy. Thus the contracting phase of the Universe would be the time reverse of the expanding phase.

In terms of arrows of time, if entropy starts to decrease during the contracting phase then the thermodynamic arrow must get flipped over (since it is defined to always point in the direction of *increasing* entropy), and if our own subjective (psychological) arrow is always aligned in the same direction as the thermodynamic one, our time will also be running in reverse. This would mean that rather than the Big Crunch being an event in our future, it would be an event in our past. Of course I am assuming that humans will survive for the billions of years necessary to put this to the test, but if we did we would not actually see the Universe contracting. Since our time would be running in reverse we would think it was still expanding. We would also therefore not see any violation of the second law of thermodynamics. After all, according to us entropy would be increasing as normal. The most fascinating conclusion to draw from this weird situation is that the Universe may in fact be collapsing at the moment, and it is only because we have an arrow of time that points in the direction of increasing entropy that we mistakenly believe it to be expanding!

I did not realize it at the time, but this idea of the reversal of the direction of time during a collapsing universe was actually due to Thomas Gold in the 1960s. Hawking tried to put the idea on a firmer theoretical footing by appealing to the quantum nature of the two singularities. In fact, the behaviour of the Universe when it is near maximum expansion would have to be very strange in Hawking's original picture. Let's say that a human survives from the expansion phase through to the contracting phase while enclosed inside a spaceship. Her arrow of time would have flipped over suddenly and she would not remember the time of maximum expansion since that would now be in her future.

I will now describe my objection to this idea. First of all, Hawking used the words 'expansion' and 'contraction' and 'surviving through the period of maximum expansion into the contraction phase'. Such language implies that there must be a separate, external, arrow of time that points from the Big Bang to the Big Crunch. Otherwise there is nothing to distinguish the two and we cannot say that one was 'before' the other. When it is claimed that we might 'mistakenly' think that we are living in the expanding phase but are 'really' in the collapsing phase, we would need such an external time to act as an adjudicator and tell us what the Universe is really doing. We know of no such arrow of time and to suggest that one might exist is reminiscent of my earlier discussion of a hypothetical external time against which we would need to measure the rate of flow of our time. And if there is no preferred overall direction of time that would label the expansion and contraction phases then the Big Crunch really should be equivalent to the Big Bang and would also mark a beginning of time. We would therefore have time flowing from both singularities, in opposite directions, towards an 'end of time' at maximum expansion.

I will highlight this problem of an end to time by considering the fate of the surviving human in a spaceship near the time of maximum expansion. She calculates that the Universe will reach maximum expansion at three o'clock that afternoon (let's call it T-max). She is aware that her arrow of time is about to flip over. At one second to three everything is normal and she knows she has a second to go. What will be happening two seconds later? It is now one second past three and we are in the contracting phase. If her arrow of time has now reversed and all processes inside the spaceship are running backwards then her clock will now say one second before three again. She will still think that the Universe has another second's worth of expansion.

Even at one millionth of a second to three on this side of T-max there would be nothing unusual, but two millionths of a second later she would still believe T-max to be a millionth of a second away. We could get as close to T-max as we liked but there would never be a time later than it. It would really mark an end to time.

The above objections do not prove that Hawking was wrong, rather that the language used presupposed an extra arrow of time that did not change directions at T-max, and to which no reference had been made.

After discussing his theory with colleagues, Hawking soon realized that the Universe need not return to a state of low entropy at the Big Crunch and hence there would not have to be a reversal in the direction of our arrow of time. The entropy of the Universe

133

could carry on increasing from the expanding phase through to the contracting phase. Unfortunately Hawking caught pneumonia and was unable to write a quick follow-up paper explaining his mistake. I vividly remember reading his best seller *A Brief History of Time* while on the train to work a year or two after it came out—a friend had bought me the paperback edition at New Delhi airport long before it was available in Britain—and I remember feeling both surprised and full of admiration for Hawking's honesty. Above all, I remember being embarrassed that the stupid grin on my face had attracted the attention of the other commuters.

So how can we understand the difference between the low entropy Big Bang and the high entropy Big Crunch? One explanation is that space near the two singularities has different geometries. Current thinking is that black holes are reservoirs of entropy. The bigger they are, the higher their entropy. Since the Big Crunch can be considered as the ultimate black hole which has swallowed the whole Universe, it should have an extremely high entropy. The Big Bang, in contrast, is like a white hole and would have very low entropy.

This is rather unsatisfying though. After all, where does gravity come in? Where does expansion come in? And how did the Universe get to be in such a low entropy state in the first place?

At first sight, it would appear that the Universe is in a state of low entropy at the moment. Stars are hot spots in space which are radiating their heat into their surroundings and causing entropy to increase (remember the idea of heat transfer was one way of defining entropy). When a star stops shining it will have completely wound down and would be in a state of high entropy (whether or not it ends up as a black hole). So there will be a time in the distant future when all the stars will have burnt out and their radiation would be spread out evenly in space (high entropy). There is a serious problem here, however, which physicists have attempted to wriggle their way out of with lesser and greater degrees of success. Before stars and galaxies had formed in the early Universe, before even matter had had a chance to form from pure energy, the Universe would have been in a state of thermal equilibrium, with its energy spread out evenly so that no one region of space was any hotter than any other. Surely this is a state of maximum entropy! So what caused the stars to form in the first place?

One proposal goes like this: It is true the Universe started off in a state of maximum entropy, but it was also very small then. The entropy it had was the maximum possible for that sized universe. The Universe then went through a period of rapid expansion (inflation) and the maximum amount of entropy it *could* have increased dramatically. However, its actual entropy quickly fell behind this maximum possible value creating an 'entropy gap'.

In his book *The Emperor's New Mind*, Roger Penrose criticizes this view by claiming that the time reverse situation should also apply if and when the Universe were to finally collapse to a Big Crunch. As it shrinks, the entropy gap will decrease until it again reaches a size in which the entropy is the maximum possible. Any further shrinking would squeeze the entropy down further, in violation of the second law.

How can we therefore understand this asymmetry between the two singularities? Can gravity provide it? An obvious difference between the expanding and contracting phases is that in the former there would have been some initial conditions that set the Universe expanding in the first place. The contracting phase on the other hand is due entirely to the gravitational pull of the matter within the Universe. Thus the physical origins of the expansion and contraction are different. But it would be satisfying to be able to explain the evolution of the Universe in terms of entropy.

Another oft-quoted difference is that a very old contracting universe would no longer have any stars still burning. It would consist entirely of cold background radiation, dead stars and black holes. Clearly a high entropy landscape. But this is not the only possible scenario. Let us for simplicity assume that the contracting Universe contains only low energy light (photons) and black holes. Hawking has shown that black holes evaporate and we can therefore imagine a universe that is so old—one that has just enough matter to close it means it would take gravity a very long time to halt and reverse the expansion—that all the black holes could have evaporated away. Whether they leave behind them empty, naked, singularities is unclear, but if they do not then the Universe will finally consist entirely of cold radiation.

A possible solution

I have still not explained how stars and galaxies could have formed in the first place. This could only have happened if there had been irregularities, or wrinkles, in the fabric of space that would cause the matter there to be more dense than average. As long as space does not expand too fast it is now inevitable that the matter in those regions will clump together further. This is similar to the example of the molecules of air in the box that I discussed earlier. In that case the volume inside the box did not expand, and the regions of slightly higher density arose by pure chance. In the early Universe, those regions where matter was clumping together would have eventually heated up so much that nuclear fusion would have been triggered and stars were born. However, the amount of wrinkling had to be just right. If too little, matter would never have clumped together and galaxies and stars (and hence us) would never have formed. On the other hand if space had been too crumpled then the high density of matter in those regions would have quickly resulted in the formation of huge black holes.

Even if we do not understand the origin of these irregularities we should at least look for experimental evidence that they existed in the early Universe. It was predicted theoretically that they should show up as tiny temperature variations in the microwave background radiation which, as I mentioned in Chapter 3, is the afterglow of the Big Bang. This effect had to be so small however that it could not be detected from Earth. In 1992, NASA announced that the COBE satellite (which stands for COsmic Background Explorer) had detected a difference in the temperature of the background radiation of just the right magnitude. The discovery was hailed as the final proof that the Big Bang model was correct. Some astronomers, however, argue that this statement is too strong and that the COBE result did nothing more than support our notions of galaxy formation.

Does everything fit together now? Did the entropy of the Universe start from a low value at the Big Bang? Will it keep increasing even if the Universe one day collapses to a Big Crunch, and hence provide us with an arrow of time that does not flip over? I believe so, assuming of course that the Universe will one day collapse (not likely, I know).

Just after the Big Bang, the Universe was hot and energetic and thus in a state of low entropy. As it expanded it cooled. Its entropy increased rapidly, not due to any heat transfer but rather because its energy can be thought of as being used up to provide the work for the expansion.

As the Universe cooled, a tiny amount of its energy became locked up inside hydrogen atoms. Then, thanks to the wrinkles in space which provided the seeds for the formation of stars, gravity was soon able to cause these atoms to clump together to form the galaxies and the stars within them. It then provided the means for tapping this energy within the atoms through nuclear fusion.

If galaxies and stars had not formed, the Universe would have died a heat death long ago. It would now be a cold black place. The energy locked up within stars is just delaying the inevitable. In a sense, the heat death of the Universe has already taken place. The galaxies are really only small isolated pockets of resistance to the rapidly increasing entropy around them. The microwave background radiation with its temperature of just three degrees above absolute zero is proof that the Universe has almost completely unwound already.

Some authors have claimed that the heat death of the Universe will never happen even if it continues to expand forever. Since the space available for the matter in the Universe is always increasing, they argue, there will always be more room for it to spread into. This is wrong. Once matter and radiation are uniformly spread throughout space then any further expansion will just reduce the density (amount of matter in a given volume). It will not alter the state of equilibrium.

If the Universe is destined to collapse again under its own gravity then this would represent a further increase in entropy. It does not matter if all it contains by then is cold radiation because no gravitational clumping in the usual sense is necessary. The Big Crunch is not like the formation of galaxies in the early Universe. During the collapse the whole Universe is closing in on itself. The best way to describe this is to think of the Universe as a spring. The expansion is like the stretching spring. If stretched too hard it will never return to its original coiled state. If the stretching is more gentle then it will allow itself to be pulled so far before it snaps back into position. In the same way, the Universe at maximum expansion still has gravitational potential energy. As it collapses its entropy increases still further. The maximum entropy is reached at the Big Crunch which marks the end of time; the tip of the thermodynamic arrow of time.

The above explanation is an over-simplification. I have mentioned that there is no real consensus yet on the arrows of time in cosmology and the reasoning I have offered will be far from the last word on the subject.

Now that you have seen just how confusing time can be as a concept on its own, you are finally ready to meet Einstein's special theory of relativity in which he managed to lump time together with space to form four-dimensional spacetime. Don't be too alarmed. Compared with this chapter's often surreal metaphysical ramblings, special relativity should be a breath of fresh air.

I sense scepticism.

6

EINSTEIN'S TIME

*"Ah! that accounts for it," said the Hatter. "He won't stand beating.
Now if you only kept on good terms with him, he'd do almost anything
you liked with the clock... you could keep it to half-past one as long
as you liked."*

Lewis Carroll, *Alice's Adventures in Wonderland*

What is so special about special relativity?

In a way, this chapter is the engine room of the book. So far, I have
asked you to imagine higher dimensions, expected you to accept
that gravity can warp space and time and to take my word for what
we think it would be like to fall into a black hole. However, I have
not covered them in sufficient depth for you to fully appreciate
the logical reasoning that led to them, as that would have been
beyond the scope of this book. This chapter is different. I cannot
brush aside the reasoning that led us to the view of space and time
that Einstein has shown us. Here is where you will see his true
genius and, I hope, appreciate the unavoidable, yet astonishing,
conclusions he was forced to reach.

Ten years before his general theory of relativity of 1915
Einstein showed, through logical necessity, how time and space
are related. This, as we shall see, is where the idea of time as the
fourth dimension comes in. It became known as his special theory
of relativity, and it was only after he had understood the structure
of this 'spacetime' that he could turn his attention to his general
theory in which he showed how gravity could curve it.

139

Einstein announced his special theory of relativity (now known simply as special relativity) to the world in 1905 while still in his mid-twenties. But he had been struggling with the concepts leading up to it since his mid-teens. Special relativity is the reason Einstein is famous today, despite the fact that it was superseded by the much grander general relativity a few years later, and that it was, in fact, the experimental confirmation of general relativity that turned him into a household name. Einstein's paper on special relativity was not even deemed to be his most important piece of work in the year it was published. Its impact took time to sink in. Remember he received the recognition of the Nobel prize committee for his work proving that light consists of particles. So what is it about the special theory that makes it so special?

Popular accounts of special relativity will often try and fob you off with the explanation that it was the theory that gave us the famous equation

$$E = mc^2.$$

This is true, and it was this simple formula which led us, for better or worse, into the nuclear age. However, special relativity goes much deeper than that. It is a bit like describing the industrial revolution as having given us the steam engine. In reality, the industrial revolution meant much more than a single invention. Not only did political power shift from the landowner to the industrial capitalist, but with the later development of the internal combustion engine and electricity came a complete change in ordinary people's lives. In a similar way, special relativity is about much more than $E = mc^2$. It heralded a revolution in physics. It showed how and why the old notions of space and time had to be ditched and replaced with such a new and unfamiliar set of concepts that, to this day, we still have not been able to shake off the 'old notions'. The space and time that most people still take for granted as 'common sense' were shown to be wrong by Einstein. Since then every experiment ever devised has only served to confirm, with ever-increasing accuracy, that he was right. We will see in this chapter why it has been so hard for many people to accept his ideas, even almost a hundred years later.

Newton is rightly acknowledged as having sewn up the whole of classical mechanics with his laws of motion. These describe how objects move and how forces such as gravity affect them by making them speed up, slow down or change direction. The most familiar of these laws is probably the third one. You probably remember it as the one about every action having an equal and opposite reaction. However, it is the second law which is the most important and fundamental—it is pure coincidence that the most important law in the field of thermodynamics is also the second one—and describes how a body will behave when pushed.

All moving objects can be divided into two categories: those that do not feel any force, and are therefore either stationary or coasting along in a straight line at a constant speed, and those which are under the influence of some force that is causing them to change either their speed or direction. Examples of the second category include falling objects, an accelerating or braking car, a car going round a corner, even a ball rolling along a flat surface since wind resistance and friction are both forces that act to slow the ball down. Newton's laws of motion cover all the above cases with an accuracy that in most everyday situations is very impressive.

Einstein's theories of relativity go far deeper than merely stating laws of motion. The reason he needed two theories was because he had to distinguish between the above two categories of motion. Bodies moving freely at constant velocities and in the absence of gravity are described by special relativity. Once the force of gravity is switched on we must turn to general relativity.

You have already seen how Newton's law of gravity is only an approximation to the more exact general relativity, but it nevertheless works very well in weak gravitational fields, such as the Earth's. In the same way, Newton's laws of motion are only approximations to special relativity, but the differences now only show up when objects move at very high speeds. For most purposes in everyday life the accuracy of Newtonian mechanics is as much as we need. Even NASA uses Newton's laws to calculate the path a rocket should take to reach the Moon, and rockets are probably the fastest moving objects most people can think of. Clearly the high speeds I am referring to, at which Newton's laws break down, are much higher than the speeds attained by today's rockets. In fact, it is only for bodies moving at a substantial fraction of the speed of light (which stands at three hundred thousand kilometres *per second*) that special relativity is required. In the following discussion I will often use examples of objects moving at close to the speed of light. This is just to highlight the effects of relativity more clearly and you should not take these examples too literally.

There are several ways that special relativity is traditionally explained. The usual way is by deriving a set of algebraic equations called the Lorentz transformation equations. Don't worry, we will not follow that route here. The second way is by using special kinds of graph called spacetime diagrams. Many authors of non-technical books on relativity use such diagrams because they feel that they are simpler to interpret than abstract equations. In a way this is true. Most people are used to seeing graphs of one sort or another. Newspapers and television show the varying fortunes of political parties in opinion polls or the fluctuations of share prices on the stock market. Most companies present data in their annual reports using bar charts, pie charts and histograms. Such graphical methods may well be informative and simple to interpret. But spacetime diagrams are another matter. If you are mathematically inclined you will most likely find them very helpful. If you are not then they will be almost as impenetrable as algebraic formulae. I will therefore adopt the third route for explaining Einstein's ideas: I will restrict myself to words only.

So what is all the fuss about? You might be wondering why I don't just get on and explain it instead of this tedious fanfare. But special relativity deserves respect. Its conclusions provide the stock-in-trade for so much of science fiction, and are synonymous with the name of Einstein. As an example I will quote two of the most frequently asked questions in the whole of modern physics. Both are direct results of special relativity. They are:

- Why can nothing travel faster than the speed of light?
- Why do clocks tick more slowly when they are moving very fast? (This has nothing to do with alarm clocks being hurled across bedrooms.)

When I am asked these questions my usual reply is that the questioner really needs to take a course in special relativity if they wish to get to the bottom of things. For there are a number of logical steps that you will need to work through before you can feel convinced. In this chapter I will lead you through those steps. If you are not interested in the answers to these questions and are happy to accept them, for it is quite true that nothing could ever go faster than light and we really do see fast moving clocks slow down, then you can skip the next few pages, but since you have reached this far I have every faith in your continued perseverance.

The two faces of light

Recall that I discussed the strange properties of light at the beginning of Chapter 4 as an introduction to black holes. I now need to return to the subject of light, not because it is merely 'useful' but because it underpins the whole of special relativity.

By the late nineteenth century, Thomas Young (the Englishman who proved Newton wrong about light being made up of particles) and James Clerk Maxwell (the Scotsman who discovered that light was made up of electromagnetic waves) had shown beyond any doubt that light behaves like waves. Today there are numerous experiments that clearly and beautifully reveal the wave nature of light. It is true that quantum mechanics has since shown that light can, under certain circumstances, also behave like a stream of particles, but for the following discussion it is its wavelike nature which we require.

An important property of waves is that they need something to move through; a medium through which the vibrations can propagate. When you speak to someone standing next to you, the sound waves that travel from your mouth to his ear need the air in between to move through. Likewise, water waves on the surface of the sea need the water, and the 'bump' that travels along a length of rope when it is given a flick at one end needs the rope. Clearly, without the medium to carry the wave along there would be no wave. This was why nineteenth century physicists were convinced that light, having been confirmed as a wave, also needed a medium. And since no one had seen such a medium, they had to think of a way of proving its existence.

It was known as the luminiferous[1] ether—not to be confused with the organic chemical used as an anaesthetic—and the hunt was on to find it. If it existed then it had to have certain properties. For a start, it had to permeate the whole of space in order for starlight to be able to reach us. It had to exist everywhere, even in the empty space inside atoms. An important property of the ether was that it could not interact with material objects and therefore could not be dragged along with them when they moved. This had been confirmed as long ago as 1729 due to a property of light known as aberration.

Nothing else was known about the ether. It was hoped that much more would become clear with advances being made in the field of optics. However, nobody was prepared for what was to come next.

In 1907, A A Michelson became the first American to win the Nobel prize in physics for an experiment he had carried out in the 1880s together with E W Morley. It is probably the most famous experiment in the whole of physics. Michelson had invented a device known as an interferometer which relies on the wave nature of light to measure the time it takes for a light beam to cover a fixed distance. By clever use of his interferometer to measure how fast light beams travel he was able to prove beyond any doubt that the ether could not exist!

An important fact in physics is that all waves travel at a speed that does not depend on the speed of the source of the waves. Think of the sound of an approaching fast car. The sound waves will reach your ear before the car since they are travelling faster, but their speed is to do with how quickly the vibrating air molecules can transmit them. They do not reach you any quicker by virtue of being 'pushed' along by the moving car. What happens instead is that the waves get squashed up to higher frequency and shorter

[1] Meaning transmitting light.

144

wavelength in front of the car (the Doppler effect) but the speed of the sound itself doesn't change[2].

Sound waves travel through air at a speed of 1200 kilometres per hour. This speed is independent of how fast the car is moving. If the car is travelling at 100 kilometres per hour then the driver would see the sound waves (assuming sound waves could be seen) moving ahead of him at a speed of only 1100 kilometres per hour (1200 minus 100). The faster the car goes the slower the relative speed of the sound waves that the driver sees. But to a stationary observer watching the approaching car, the sound waves always travel at 1200 kilometres per hour no matter how fast the car is moving. If the driver and the stationary observer had an argument about the speed of sound, the driver would have to admit that the speed he observes the waves moving at is not their true speed because he too is moving relative to the air molecules.

Michelson and Morley applied this principle to light waves. They assumed that the Earth is moving through the ether as it orbits the Sun (at about a hundred thousand kilometres per hour). Their experiment is a little tricky to describe so I will not go into the finer details. Suffice it to say that they measured with very high accuracy the time it took light in a laboratory to travel along two paths of equal distance, one in the direction of the Earth's motion as it orbited the Sun and the other at right angles to it. Sitting in their laboratory on Earth and observing the speed of light they were like the car driver who would measure the sound waves leaving the car at different speeds depending on what direction he looked. After all, to him the sound waves that were travelling straight upwards would still be moving at 1200 kilometres per hour.

If the ether existed, and Michelson and Morley knew that the Earth had to be moving freely through it, then the light moving along the different paths would cover the two equal distances in different times. This would indicate that, relative to the moving Earth, the light was moving at different speeds in the two directions. Although the speed of light is three hundred

---

[2] The speed of a wave is obtained by multiplying its frequency by its wavelength. If one quantity increases while the other decreases then they will still balance to give the same answer.

thousand kilometres per second, which is ten thousand times faster than the speed of the Earth, Michelson's interferometer was still accurate enough to pick up any difference in the timing between the two beams if there was any. None was found. Many more precise experiments using laser beams have since then confirmed Michelson and Morley's result.

Their experiment had shown that light was not like other waves. It travels at a speed that is the same whether you are moving towards the source or away from it. It doesn't have a fixed background against which its speed can be measured. So there was no need for the ether at all.

Most physicists at the time refused to believe this and tried to modify the laws of physics to accommodate the new result but to no avail. They tried to argue that light was behaving as a stream of particles (since that would also explain the result) but the experiment was set up specifically to detect the wave nature of light. It detected intereference patterns between the waves in a manner quite similar to Thomas Young's original set-up which confirmed the wave nature of light in the first place. In any case, light behaving as particles would also do away with the need for an ether since they would not require a medium to travel through.

Thought experiments and brain teasers

Einstein was only a child when Michelson and Morley carried out their experiment. Even during his youth he pondered the unusual properties of light by devising thought experiments (his famous *gedanken*). He tried to imagine himself flying at the speed of light while holding a mirror in front of him. Would he see his own reflection? How could the light from his face ever reach the mirror if the mirror itself was moving away at the speed of light? His years of contemplation culminated in two simple statements known as the principles of relativity. They can be put in the following way:

1.   There are no experiments you could perform that would tell you whether you were standing still or moving at constant

speed. All motion is relative so nothing can be said to be truly stationary.

2. Light behaves like a wave in that its speed does not depend on the speed of its source. At the same time it does not require a medium to travel through like other waves.

So far, so good. You would think that there is nothing in the above innocuous statements that you might have difficulty subscribing to. They certainly look too lightweight to be able to answer the two questions posed earlier in the chapter about nothing going faster than light and time slowing down. They may look harmless but believe me, by accepting them you will be selling your soul to the devil.

First let me assure you that they are both quite true and can be demonstrated easily. The first postulate suggests that if you perform a simple experiment like dropping a ball while on board an aircraft travelling at a constant speed, the ball will, according to you, fall vertically in the same way that it would if you performed the experiment on the ground. You therefore have just as much right in claiming that the aircraft is stationary while the Earth is moving beneath you at several hundred kilometres an hour in the opposite direction. A clearer example is that of two rockets travelling at constant speeds towards each other in space. If both the rockets' engines are off and they are just 'cruising' they could never decide whether they were both moving towards each other or whether one was stationary and the other approaching it. It is no good appealing to a nearby star as a reference point since who is to say that it is really stationary?

The second postulate was confirmed by the Michleson and Morley experiment and on its own seems harmless. It is when the two postulates are combined that the trouble starts. I know I sound a bit like a doctor, but I want you to be brave as this might hurt a little.

We have established that the light reaching us from a source will travel at the same speed regardless of how fast the source is moving. But because it doesn't have a medium to travel through and with respect to which we can measure its speed, then we can equally well say that it is not the source moving towards us but us moving towards the source since all motion is relative. This is just a statement that light obeys the first principle of relativity.

Consider the two rockets approaching each other again. An astronaut aboard one of them shines a light beam towards the other and measures the speed of the light as it leaves her rocket. Since she can quite legitimately claim to be stationary, with the other rocket is doing all the moving, she sees the light moving away from her at the usual three hundred thousand kilometres per second. At the same time, the astronaut aboard the other rocket can also legitimately claim to be stationary. He will measure the speed of the light reaching him to be three hundred thousand kilometres per second and states that this is not at all surprising since the beam's speed does not depend on how fast its source is approaching.

Both measure the light to have the same speed. This is amazing, and goes quite against common sense. Both astronauts measure the same light beam to be travelling at the same speed, despite moving relative to each other!

We can now answer Einstein's question involving the mirror. It does not matter how close he gets to the speed of light—and I will explain later why he could never travel *at* the speed of light—he will always see his reflection. This is because regardless of his speed he still sees light travelling at the same speed from his face to the mirror and back again.

A better way of formulating this is to imagine shining a torch, then travelling alongside the torch beam at three quarters of the speed of light according to someone left holding the torch. Your common sense tells you that you should see the light still overtaking you but at a much reduced speed of one quarter its original speed. Right?

Wrong! You still see it moving at the same speed that the person holding the torch measures it to be moving.

Slowing down time

We have reached the above strange state of affairs by following a number of logical steps coupled with firm experimental findings.

So where are we going wrong? After all, it is the same light beam; the same electromagnetic waves or photons or whatever you choose the light to consist of, that is leaving the torch. How can you, while travelling alongside it at some healthy fraction of the speed of light, still see it moving past you at the same speed as that seen by the guy holding the torch? The only way this could happen is *if your time is running at a slower rate than his.* If he could see a stopwatch you are holding he would see it counting by the seconds more slowly than his. If he could somehow remotely measure your heartbeat he would find it slower. Everything about you is, according to him, running slower. That's not all; if you forget about the light beam for a moment, the first principle of relativity implies that you could equally well consider your friend who is standing on the ground to be the one who is moving at three quarters the speed of light, in the opposite direction. You would see his time running slower than yours!

This is not some crackpot theory devised to make sense of the ridiculous notion that light would travel at the same speed for everyone. The notion about the speed of light is far from ridiculous and is confirmed all the time these days in experiments in particle accelerators. These are huge laboratories with circular underground tunnels, several miles long, that send subatomic particles round at close to the speed of light, such as the famous CERN facility in Switzerland. The slowing down (called dilation) of time is an unavoidable consequence of the behaviour of high speed particles.

Let me first quickly mention these particle experiments. It is known that a certain type of subatomic particle, called pions (pronounced 'pie-on'), emit photons of light. When a pion is stationary the photon will, of course, emerge at the speed of light (it is a particle of light after all). But at CERN, pions can be made to move round in a large circular underground tunnel at very close to the speed of light. They still emit their photons however, and those photons emerging in the direction that the pions are moving can be detected and their speed measured. They are found to be still travelling at the same speed that they travel when emitted from a stationary pion.

Thus the same photon emerging from the moving pion is seen to travel at the speed of light from our point of view standing in the laboratory and from the point of view of the pion itself.

As for the slowing down of time, we can see how this comes about by considering the following thought experiment. Figure 6.1 shows a box containing a light source with a detector at the bottom and a mirror at the top. The source, which is pointing straight up, emits a flash of light (called a light pulse) which bounces off the mirror at the top and back down into a detector which signals when it has received it. According to someone inside the box the light will take a certain time to go from the source to the mirror and back to the detector. Now imagine that the whole box is itself moving sideways at close to the speed of light. To an observer watching it zoom by (it has a glass front), the light pulse traces a path that is longer than the straight-up-and-down path seen by the person inside the box. To the observer watching from outside, the pulse must cover the longer distance shown in the figure, but he still sees the light travelling at the same speed. However, since it must now cover a longer distance (the dashed line), a longer time will have elapsed before it gets back to the detector[3]. Therefore more time goes by according to a clock on the ground than according to a clock inside the box. Since both clocks are measuring the duration of the same process (the time taken for the light to move up and down the box) time inside the box must be running slower for its clock to record a shorter duration! Aficionados of special relativity will be aware that this explanation is not strictly the whole story since to say that someone 'sees' something implies that light must reach that person's eyes from the object, and it will take a finite time to reach them.

So moving clocks run slow and the above example shows how that happens. Often, when people encounter this effect for the first time, they have the impression that the rapid motion affects the mechanics of the clock; that the clock is somehow responding to the speed at which it is moving. This is quite wrong. In fact, since

---

[3] The time taken is given by the distance the light pulse has travelled divided by its speed. This, I hope I don't need to remind you, comes simply from the relation: speed equals distance over time. So the further it has to go, the longer it will take.

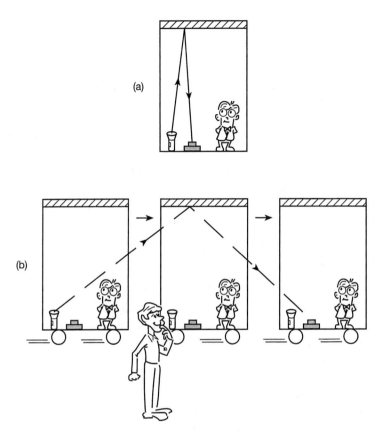

Figure 6.1. (a) An observer inside the moving box sees the light pulse cover a distance that is roughly twice the height of the box in its there-and-back journey. (b) To an outside observer, the light covers a longer distance. The three boxes are snapshots of successive timeframes. The left hand one is the position of the box when the light pulse is emitted, the middle one is where it is when the pulse reaches the mirror and the right hand one is where it is when the pulse reaches the detector. If both observers agree on the speed of the light (as they must) then the only way it can cover different distance is if they disagree on the time it takes to complete its there-and-back journey.

all motion is relative, the person inside the box in the last example can rightly claim not to be moving at all and that it is the outside observer who is travelling at close to light speed. This is borne out by the fact that he will indeed see the clock on the ground running slower than the one inside the box! This often gives rise to what appears to be a logical contradiction. How can both clocks be running slower than each other? People who do not understand relativity surmise that the clocks only *seem* to run slow according to each other because it takes light a certain time to travel from the clocks to the other observers. As James A Smith states in his book *Introduction to Special Relativity* "nothing could malign the theory of relativity more thoroughly". We will see in the discussion of the paradox of the twins later on how we can slow time down permanently by making a clock speed up and slow down again.

I am sure you must be thinking that this is, after all, just a theory. It may be fine for science fiction writers but surely it can have no place in the 'real world'. If the rate at which clocks tick can be so dependent on their relative motion, why would we bother about such things as high precision timekeepers like atomic clocks? The reason is that the effect only shows up when clocks are travelling at extremely high speeds relative to each other. The closer to the speed of light that a clock moves, the slower it will tick. If it were to travel at the speed of light relative to us then we would see its time stand completely still.

Here is another example. Consider a sprinter who runs the hundred metres in exactly ten seconds, according to the reliable and highly accurate timekeeping of the judges. If he had carried his own very accurate stopwatch with him then, due to time slowing down very slightly for him while he was running, his watch would show a time of 9.999999999999995 seconds. Of course, this is so close to ten seconds that we would never know the difference. However, scientists routinely need to measure times with this sort of accuracy. The difference between the runner's and the judges' watches is just five 'femtoseconds'. The reason it is such a small time difference is because the athlete is moving much more slowly than light. Even the fastest rockets are too slow to show an appreciable effect.

Can we therefore ever see real time dilation in action? Well, this is something I can vouch for personally because, like many physics students, I performed a laboratory experiment while I was a student at university. The experiment involves another type of subatomic particle called a muon ('mew-on') which is produced by cosmic rays. These are high energy particles from space that are continually bombarding the Earth's atmosphere. In the upper atmosphere many new types of particle, mostly muons, are created in this way, and travel down to the surface of the Earth. Physicists have studied the properties of muons and know that they have an extremely short lifetime of one millionth of a second. This lifetime is, of course, only statistical in that some muons might live for a little longer, some for a little less. But if a thousand muons are created at once then after a millionth of a second there will be roughly five hundred left.

The muons created in the upper atmosphere are so energetic that they travel towards the Earth at an incredible 99% of the speed of light. However, even at this speed it should still take them several lifetimes to cover the distance to the surface of the Earth (and, more importantly, into the muon detector in the laboratory). We should therefore observe only those few with unusually long lifetimes that were able to complete the journey. Instead we find nearly all the muons are comfortably able to complete the journey. The reason this is possible is that the muons' time (their internal clocks that measure their lifetime) is running much slower than ours. So from the muons' point of view only a fraction of their lifetime has elapsed.

An alternative argument that fast moving muons must for some reason live longer than stationary ones does not hold water. On closer scrutiny we see that this cannot be correct since it would be violating the first principle of relativity: a moving muon is only 'moving' relative to us.

Shrinking distances

Not content with overthrowing the old notions of absolute time, Einstein still had a few more surprises up his sleeve. Consider how things would look to you if you were sitting on a muon as it travelled down to Earth. You would agree with someone standing on the ground watching you that you were approaching each other at 99% of the speed of light. How is it that he would see you covering the distance of one mile, say, in a time of five millionths of a second (five muon lifetimes) according to his Earth clock, while you claim to cover the same distance in just one tenth of that time. There are no light beams involved here and you would think that the only maths required is the relation: speed equals distance over time. How is it that both of you agree on the speed you are moving and yet cannot agree on the time it takes you to cover the same distance?

Something else has to give, and now it is distance. In order to obtain the same value for the muon's speed in both cases (by dividing distance over time) the distance travelled as seen by the muon must also be one tenth of its value as seen from Earth. That is, the muons will see the distance squashed up to much less than a mile. This explains how it is able to survive the journey; it does not think it has had so far to travel.

This property of high speed travel is known as length contraction. It states that fast moving objects look shorter than they do when standing still. In the example of the muons the object in question is the thickness of the atmosphere. An Irishman and a Dutchman first suggested this effect soon after Michelson and Morley's experiment, and several years before special relativity. George Fitzgerald and Hendrik Lorentz pointed out that the result of the ether experiment could be explained if there was a contraction of lengths with high speed motion. This would have rescued the idea of the ether. Lorentz even went so far as to derive a set of equations that now bear his name. Unfortunately for him, he did not make that final leap of intuition that was the second postulate of relativity. In a way, therefore, a lot of the groundwork had already been done before Einstein and it is often claimed that, had he not discovered special relativity, someone else would have.

As with the slowing down of time, the shortening of lengths is something which shows up more the closer a moving object gets to the speed of light. So what sort of effect would we observe in the real world? To give you a solid example, imagine taking a high precision photograph of a jet that is flying at twice the speed of sound (over two thousand kilometres per hour). You would observe it to be ever so slightly shorter than it was when on the ground. But for a typical aircraft this shortening of length would be less than the width of a single atom! This is certainly not measurable from a photograph of the aircraft. But you have to remember that although twice the speed of sound seems impressively fast to us, it is nothing compared with the speed of light. If the jet had been travelling several hundred thousand times faster, say over three-quarters the speed of light, then we would see a difference. The jet would look only half its original length. If it were to travel as fast as the cosmic ray muons, it would look squashed to just one tenth of its length.

How uncomfortable for the poor pilot, you must be thinking. Presumably this is one of the hazards of such high speed travel. The truth is that the pilot will feel nothing unusual. To him the dimensions of the plane (and himself) have not changed at all. In fact, due to the first principle of relativity, he sees the world around him squashed up, in the same way that the muons would (if they could see that is!)

Light—the world speed record

There is nothing that annoys people more about relativity when they first encounter it than the claim that nothing can travel faster than light. They are prepared to accept clocks slowing down, lengths shrinking, even that light travels at the same speed for all observers, but why in heaven's name can we not conceive of anything moving at a speed of over three hundred thousand kilometres per second? Granted, this is a stupendously high speed to which nothing that we know of (apart from subatomic particles) can get close, but special relativity seems to be saying that the laws of nature *forbid* anything from going faster. Imagine building a rocket that could keep accelerating faster and faster. Of course, such a machine is way beyond our current technological ability, so what if an alien civilization were to build it? What will happen as it reaches the speed of light? Does some cosmic speed ramp become activated? Does the rocket blow up, fall into a black hole or enter a time warp? Nope, nothing so dramatic.

There are a number of ways to explain why the speed of light is the upper speed possible in our Universe. One method is by using algebra. (Oh great, you're thinking, that will really convince me.) However, I will not go into the gory details. Suffice it to say that, in special relativity, speeds get added up in a very strange way. If you are on a train moving along at 100 kilometres per hour and you throw a ball out of the window at ten kilometres per hour in the direction the train is moving then, to someone standing outside watching you go by, the ball will initially (before the wind has slowed it down) be moving at a combined speed of 110 kilometres per hour. This is known as the law of addition of velocities. What if we now restate the same example but with much higher speeds? Consider what the outside observer sees in figure 6.2. The rocket is moving at three-quarters the speed of light when it fires a missile that flies off at half the speed of light as seen by the man in the rocket. Does the observer see the missile moving at one and a quarter times the speed of light? He would if the usual rule about adding velocities were correct. But like so much of physics that is valid for everyday use, this law breaks down at relativistic speeds. The correct formula to use would say that the observer sees the missile moving at nine tenths of the speed of light. It does not matter how close to the speed of light the rocket and missile were moving, their combined speed according to the stationary observer would always be greater than their individual speeds but below the speed of light.

The easiest way of explaining the speed of light barrier also happens to be a way to explain where Einstein's most famous equation ($E = mc^2$) comes from. Once Einstein understood how space and time were affected close to the speed of light, he went on to consider what else had to be corrected. Some of the most important and basic laws in the whole of physics are known as conservation laws, which state that certain quantities should remain constant even when other quantities are changing. One

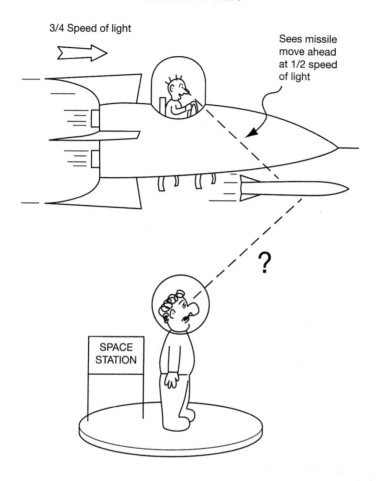

Figure 6.2. According to the normal rules of adding up speeds, the space station observer should see the missile travelling at $\frac{3}{4} + \frac{1}{2} = 1\frac{1}{4}$ times the speed of light. Einstein showed that nothing can go faster than light and the way we add up speeds had to be changed.

of these is the law of conservation of momentum. Remember the momentum of a body is given by its mass multiplied by its velocity, so a cannon ball slowly rolling along the ground can be stopped in its tracks by a bullet hitting it head-on. This will happen when the two have equal but opposite momentum which cancel each other out. The cannon ball has a large mass but low velocity, whereas the bullet has a small mass and high velocity. In both cases the product of mass and velocity can give the same number (the momentum). When any two objects collide, we expect their combined momentum before and after the collision to be the same. They don't have to cancel each other—that is a special case—but usually one will transfer some of its momentum to the other. Einstein found that when bodies travel at close to the speed of light the total momentum is not conserved, as it should be, according to some observers if they just use the simple 'mass times velocity' rule. Again, something had to give. This time it was the definition of a fast moving body's mass.

It turns out that the faster an object moves, the heavier it becomes, and the harder it gets to make it go even faster. The closer it gets to the speed of light, the larger its momentum becomes, but this is by virtue of its increasing mass, not its velocity.

Consider what happens to an object's mass when it moves very fast. One of the most important consequences of the equations of special relativity is how mass and energy are related. Einstein showed that mass can be converted into energy and vice versa. The two are related through the equation $E = mc^2$, which tells us how much energy is locked up in any given mass. The c stands for the speed of light, and thus the quantity c^2 (the speed of light times itself) is a very large number indeed and explains how we can get so much energy out of a small amount of mass. This equation suggests that we can think of mass as frozen energy.

Since a moving object also has energy due its motion (called its kinetic energy), its total energy will be the sum of the energy frozen as mass when it is not moving plus its kinetic energy. The faster it moves the more energy it has. This means that the real mass of an object will be due to its frozen energy plus the energy due to its motion. Most of the time the frozen energy of an object (its mass) is so much more than the energy of its motion that we can ignore the latter and take the mass to be the same as it was when not moving. But as the speed approaches that of light the kinetic energy becomes so great it can exceed the frozen energy.

Thus the mass of a fast moving object is much greater than its mass when stationary. Of course, as far as the object itself is concerned, it can claim to be stationary (since all motion is relative) and is thus unaware of any change in its mass.

You can now see the problem of trying to attain light speed. Imagine an accelerating train engine pulling a single carriage. What if, for every ten kilometres per hour faster that it goes, another carriage is added. It would therefore have to work harder just to maintain its speed. The faster it goes the more carriages it has to pull, and the more power it needs. In the same way, the faster a body moves, relative to some observer, the heavier it will appear, and the harder it will be to make it go any faster. To accelerate it up to the speed of light would require an infinite amount of energy, which is impossible.

When time runs backwards

Special relativity tells us that nothing can be accelerated up to a speed greater than that of light, but it does not rule out things travelling faster than the speed of light as long as they always remain on the other side of the light speed barrier. You see the speed of light is a two-way barrier; nothing moving slower than light can ever go faster than light and nothing already faster than light can slow down to a speed below that of light. Physicists even have a name for hypothetical 'superluminal' particles that travel faster than light. They are called tachyons and, if they exist, would have some strange properties. For instance, since time slows down the closer a particle gets to the speed of light, until at light speed time stands still, we can take this a step further and see that, for tachyons, time would be running backwards. To us, tachyons would be travelling backwards in time! Tachyons would not be like normal particles that slow down as they lose energy. Instead, they speed up, and when a tachyon has lost all its energy it will be travelling at infinite speed!

Even though special relativity predicts that such particles could exist, no evidence whatsoever has been found for them and most physicists do not believe they exist. Anyway, they are too weird to contemplate, even by the standards of modern physics.

There is one sure example of particles travelling faster than light though. The one thing I have not mentioned yet is that this maximum speed limit I have been discussing refers specifically to light travelling through empty space. This is called the speed of light in the vacuum. When light travels through a transparent material such as glass or water, it moves more slowly. This is what gives rise to refraction (the reason a spoon looks bent when placed in a glass of water and why a swimming pool looks shallower than it really is). Because of this, it is possible for a particle to be moving through such a medium at a speed that is greater than the speed of light through that medium. When electrons move through water faster than the speed of light in water, they emit a pretty blue light known as Čerenkov radiation. This is the light equivalent of a sonic boom when a jet breaks through the sound barrier.

Finally, there are a number of examples where it looks as though the speed of light barrier is being broken, but which on closer examination show this not to be the case. The most famous of these is known as the searchlight paradox. In Chapter 2 we met the rapidly spinning neutron stars, known as pulsars. Some of these can spin at over one hundred times per second. As they spin, they emit an intense beam of radio waves which sweeps past the Earth like a searchlight with the same frequency as the rotating pulsar. But since pulsars are so far away (usually thousands of lightyears) the spotlight that this beam shines onto the Earth must be sweeping past in such a gigantic circle that it would have to travel at trillions of times the speed of light! On closer examination, however, we can see that this is not so. The radio waves coming from the pulsar are in the form of photons (since they are just electromagnetic radiation) and it is these photons which are the things that are being emitted from the pulsar and they always travel at the speed of light. The confusion arises because we think there is something moving round in a circle, but all the photons are travelling radially outwards. No physical object is actually moving in this large circle at all. Think of the photons as being like water sprayed out of a garden sprinkler.

Little green men

Time dilation and length contraction provide a way of travelling intergalactic distances across space (which has turned out to be a godsend for ufologists). In Chapter 3 I gave you some impression of the enormous distances between the stars. Even our closest neighbours are so far away that it takes their light several years to reach us, while most stars are thousands of lightyears away. This would appear to rule out any possibility of us ever reaching other stars, along with their possible planetary systems, within a human lifetime.

This is where relativity comes to the rescue. If a spaceship could travel close to the speed of light, then it sees the distance it needs to cover contracted. A journey to a star a thousand lightyears away might only take a few years according to the astronauts on board. The catch is that, due to the effects of time dilation, the journey would still take over a thousand *Earth* years. After all, as seen from Earth, the ship is moving below light speed. On Earth we would see the ship having to cover the full uncontracted distance, but we will also see the ship's clocks slowed down so that we would also agree with them on the amount of time the journey has taken according to them.

Does this mean that space travel across the Universe is possible? In principle, yes, you could travel all the way across the Visible Universe covering billions of lightyears in one day without ever reaching the speed of light. This is despite the fact that light, which is moving faster than you, takes billions of years to make the same journey. The trick is to always make the distinction between the time the journey takes according to clocks in your rocket (that measure one day) and the time it takes as measured by clocks that remain on Earth (billions of years). A rather baffling consequence of this is that for light itself time stands still. If you could attach a clock to a light beam it would not tick at all. We say that to a photon, time does not go by at all (maximum time dilation) and the whole Universe has zero size (maximum length contraction)!!

Since it is possible to cover any distance across space in an arbitrarily short time by travelling close to the speed of light, it appears that I have, in theory at least, opened the door to the possibility of us having been visited by beings from other worlds. The argument goes that it is at least possible for a visiting alien spacecraft to have a propulsion system that is hundreds, even thousands, of years more advanced than anything we could conceive of, and could reach speeds close to that of light. They may therefore be able to cover the vast distances across space in just a few months or years. Much as I hate to pour cold water on this particular fantasy, it is highly unlikely (but not impossible of course) that UFOs are genuine flying saucers for a number of practical reasons. Since any alien spacecraft must obey the same laws of physics as everything else in the Universe, it cannot travel faster than the speed of light. Even though the journey from their home planet might only take a few years according to the travelling aliens, many thousands or even millions of years will have elapsed back on their home planet. So, assuming that their life spans are comparable to ours, it would be impossible for them to ever return and report back on their findings. Their contemporaries would all be long dead. Of course who am I to judge (a) what an alien lifetime is and (b) whether they would have any intention of returning home anyway.

Fast forward to the future

Special relativity has thrown up a number of intriguing and bizarre concepts, chief among which is the idea that time slows down for fast moving objects. One important aspect of this strange effect that I have not mentioned so far is that it gives us a way of 'fast forwarding' through time: to travel through time into the future! So let's take a closer look at this. Over the years, relativity theory has provided a rich source of debate and discussion, and not just among physicists. But by far and away the most puzzling, most debated, and yet still most misunderstood of all its consequences is known as the clock paradox, or the paradox of the twins. I shall give a brief outline of it here and show why there is really no paradox at all.

Meet the twins Alice and Bob. Alice is the adventurous one who enjoys travelling around the Galaxy in her high speed rocket, while Bob prefers to stay at home. One day, Alice bids her brother goodbye and heads off in her rocket to the *Alpha Centauri* system four lightyears away, travelling at two thirds the speed of light. Bob monitors her progress and calculates that she should reach her destination in six years' time. Once she gets there she will turn around and head straight back. Taking into account the turn-around time he expects the round trip to take a little over twelve years. He is frustrated however by the messages he receives from her. Not only is there an increasing time delay due to the widening distance between them, but they are also Doppler shifted towards longer wavelengths. From the rocket's speed he works out how much of a shift there should be and takes it into account. However, the wavelengths are still too long and he quickly realizes that this is due to the relativistic effect of time dilation. To him, time on board her rocket is running a little slower than his and this manifests itself in a longer wavelength in the signal. Taking this slowing down of the rocket's time into account, Bob calculates that according to Alice the journey should take just nine years, three years less than the duration of the journey according to Earth time. This would mean that, on her return, Alice will be three years younger than her twin brother! This is because time dilation is not something that affects only moving clocks, but all time on board the rocket, including Alice's biological clock.

This is, in fact, not the source of the 'paradox' of the title of the story. Bob has quite correctly used the equations of special relativity and computed the time difference between his clocks and his sister's. No, the paradox, or what at first sight appears to be a paradox, is that Alice does not believe her brother's predictions. She argues that the first principle of relativity is being violated here. Surely, since all motion is relative, she has just as much right in claiming that it is not her rocket that is moving away from Earth, but the Earth that is moving away from the rocket. It is Bob who is moving at two thirds light speed and it is his clocks that are running slower. She therefore claims that, on her return, she should expect her brother to be the younger of the two. This apparent symmetry has been the source of much confusion over the years. Both twins cannot be right, can they?

There are many ways of correctly resolving the problem. I will mention here the simplest one. The answer is that Bob is right and Alice is not. She will indeed return younger than her brother. Many books on relativity will state that this is because Alice is the one who must undergo acceleration and deceleration in the rocket and it is this that breaks the symmetry between the two twins. This is true, but *saying* that their situations are not the same is not *explaining* anything. The reason Alice ages less can be explained, not because of time dilation, but length contraction. To her, the distance to *Alpha Centauri* is not four lightyears but only three, and travelling at two thirds light speed means she can make the trip there in only four and a half years instead of Bob's estimate of six. A return journey of a further four and a half years means the whole trip will take nine years, just as Bob had calculated from the time dilation of her clocks. The reason why Alice gets the wrong answer by appealing to the fact that she sees Bob's clocks run slower, is that she is not using the equations of special relativity correctly. They only apply to observers who do not change their speed or direction. She does, Bob doesn't.

A time difference of three years on Alice's return may not sound very impressive so let us assume she had been travelling even faster, at say 99% of the speed of light. She would now return to Earth (if we ignore turn-around time) after eight years and one month of Earth time (which is one month longer than it would take light to complete the trip). But according to Alice, the trip would take only one year. If she had decided to travel further afield at this speed on a journey that, for her, would take ten years, then she would find on her return that eighty years had elapsed on Earth and that Bob, along with almost everyone she knew, had already died. She, on the other hand, would be just ten years older than when she left. This is a clear example of time travel into the future. If her rocket could have been nudged even closer to the speed of light she would have returned thousands, or even millions, of years into the future. So, forget Oil of Ulay. Just hop on board a fast moving rocket and zip around the solar system for a while. Friends will be amazed at how young you have kept!

It is sometimes thought that the symmetry between Alice and Bob's motion is retrieved if we consider what things look like to a third observer, say a passing space traveller. Wouldn't he see Alice and Bob flying apart and back together again? If he is moving in the same direction as Alice but at half her speed relative to Earth, he will see the twins moving away from him in opposite directions at the same speed: a symmetric picture. The problem is that Alice has to return. If the space traveller continues at the same speed in the same direction he will see Bob continue to move away from him, but Alice will eventually turn around and come towards him. She will pass him on her return journey. Thus the symmetry is broken.

For several years now I have set the twins paradox as a coursework assignment for my students at Surrey. They are asked to investigate it using different approaches. So far, the best way I have seen has been to use the third observer as an adjudicator. The maths works out very nicely.

I have come across many people who initially think that this sort of time travel into the future implies that the future must be already out there, existing alongside our present. This is not the case here. What is happening is that the future is unfolding on Earth all the time that Alice is away. It is just that, since less time elapses for her, she is moving on a different time track to Earth's.

Spacetime—the future is out there

Now that I have brought up the thorny subject of whether the future is already out there, we might as well confront it head-on and see what special relativity has taught us.

Two years after Einstein published his paper on special relativity one of his old university lecturers, Hermann Minkowski, suggested that all this business of time slowing down and distances being squashed was just a matter of different perspectives of different moving observers. But it is not the sort of perspective we are used to in 3D space, rather it is a perspective in four dimensions. Minkowski showed that time and space can no longer be treated as separate entities but are unified into what is known as spacetime. Many people, even scientists, have been confused by the need for such a picture, and it is important to appreciate why Minkowski came to this conclusion.

If you look at a solid object, such as a cube, you see that its dimension of depth (being the one in the direction of your line of sight) appears shorter than the other two dimensions of breadth and height, making the sides of the cube look squashed. Now consider someone else looking at the same cube from the side. To her, the dimension which you consider to be the cube's width is now its depth, and she sees the side facing you as squashed. The two of you will not argue about who is viewing the cube from the *correct* angle because you both understand that it is only a matter of different perspectives. Special relativity teaches us that fast moving observers must view the world within 4D spacetime, in which both spatial and temporal distances become just a matter of perspective. An observer moving at high speed relative to another will see spacetime from a different angle. According to one observer the dimension of time may look shorter or longer than it does to the other, but neither observer has a right to claim that their perspective of spacetime is more correct than any other.

Think of two separate events, say my writing this sentence and your reading it. The pre-Einstein (Newtonian) view would be that these two events are separated in space and time independently. Both the spatial distance between the place that I wrote it and the place where you read it (let's say 1000 kilometres) and the temporal distance between the times of writing and reading (say two years) are the same for all observers. Special relativity has shown how both these quantities will vary, depending on the observer. What is neat about 4D spacetime is that we can define within it a single 'distance' between the two events which is a combination of a space part and a time part. Such a spacetime 'interval' has a fixed value for all observers. So we only get back to absolute objective distances when space and time are combined.

Minkowski's 4D spacetime is often referred to as the block universe model. Once time is treated like a fourth dimension of space we can imagine the whole of space and time modelled as a four-dimensional block. To visualize this I recommend that you throw away one of the spatial dimensions (as discussed earlier in the book) so that time can take up the third dimension, represented in figure 6.3 by the axis running from left to right across the page.

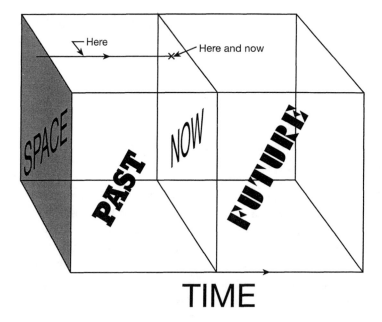

TIME

Figure 6.3. The block universe. One dimension has been discarded and space is reduced to a 2D sheet. Time runs at right angles to the sheet from left to right. If you stand still you will trace a horizontal line, called your worldline. What you think of as 'now' will be a slice through the block which includes all points in space that you consider to be simultaneous. But if two observers are moving past each other they will not agree on the same 'now' sheet.

At any given time, two-dimensional space will be a slice through the block. The Universe at earlier times is represented by the region to the left of this slice and future times to the right. Here we have a view of the totality of existence in which the whole of time—past, present and future—is laid out frozen before us. Many physicists, including Einstein later in his life, pushed this model to its logical conclusion: in 4D spacetime, nothing ever moves. All events which have ever happened or ever will happen exist together in the block universe and there is no distinction between past and future. This implies that nothing unexpected can ever happen. Not only is the future preordained but it is already out there and is as unalterably fixed as the past.

Is this picture really necessary? After all, we can just as easily imagine a Newtonian spacetime modelled as a 4D block. The difference is that in that case space and time are independent of each other, whereas in relativity the two are linked. One of the consequences of relativity is that no two observers will be able to agree on when 'now' is. By abandoning absolute time we must also admit that the notion of a universal present moment does not exist. For one observer, all events in the Universe that appear to be simultaneous can be linked together to form a certain cross sectional slice through spacetime which that observer calls 'now'. But another observer, moving relative to the first, will have a different slice that will cross the first. Some events that lie on the first observer's 'now' slice will be in the second observer's past while others will be in his future. This mind-boggling result is known as the relativity of simultaneity, and is the reason why many physicists have argued that since there is no absolute division between past and future then there can be no passage of time, since we cannot agree on where the present should be.

Worse than that, if one observer sees an event A occur *before* an event B, then it is possible for another observer to witness B *before* A[4]. If two observers cannot even agree on the order that things happen, how can we ever define an objective passage of time as a sequence of events?

Not all physicists are prepared to take such a view. Even Einstein was forced to admit that although space and time are fused into one continuum we must nevertheless not fall into the trap of treating time like an extra dimension of space. After all, we know from the last chapter that the time axis has a certain direction; an arrow of time. None of the three space axes are like this. It is just as easy to go in either direction in space, so time and space remain distinct in nature. So when Minkowski first presented his

[4] This could only happen if the time gap between the two events (in both frames) were shorter than it would take light to get from one event to the other. This then rules out the chance that one event could be the cause of the other, since this would imply that one observer would see the effect happen before the cause, and the laws of physics will not allow such a thing.

ideas, even Einstein was sceptical. He came round slowly to the idea though, and it proved vital for his subsequent development of general relativity, in which it is 4D spacetime that is affected by gravity. In fact, spacetime can be curved, stretched, squeezed and twisted. We shall even see later on that general relativity allows spacetime to form some very strange shapes known to the experts as 'non-trivial topologies'.

Where does all this leave us? Is relativity asking too much? We know that we must give up the idea of a universal present moment, but are we forced to concede that the future already exists as well? I will put forward three reasons as to why I do not believe this to be the case.

Firstly, the disagreement that any two observers will have over the ordering of events will only involve those events that are very close together in time. Imagine two flashes of light that according to me are separated by thirty centimetres. It would take light one billionth of a second to cover this distance and so, in order for the light of one flash to have triggered the other one, I would have to see them separated in time by over a billionth of a second. With such a time gap, it would have been impossible for any other observers to have seen the light flashes happen the other way round, however fast they might be moving relative to me. This would violate a sacred law of nature that states that the cause of something must always happen before its effect. It stands to reason that we cannot have things happening *before* whatever caused them to happen in the first place. In this example it does not matter whether or not the first flash of light caused the second, simply that there was enough time for this to be possible.

Therefore the reordering of events for different observers is only allowed—if it is not to violate what is known as 'causality' (or causes happening before their effects)—if the two events are so close in time that no signal, not even light, could have passed between them[5]. The mixing of the order of events is thus something that ruins the objective passage of time only on a very

[5] Of course, the events could be separated by a whole year as long as they are more then a lightyear's distance apart (since it would take light, the fastest means of signal transmission, more than a year to get from one event to the other).

small scale, which makes the present moment a little 'fuzzy', that is all.

The second reason is that, whatever your relative state of motion, there is still a definite 'now' and hence a perfectly sensible split of events, for you, into past and future.

Thirdly, and as for the future being 'already out there', it is clear that until it has 'happened' for us and we know the whole of spacetime we cannot cut slices through it anyway. To us, the future has not happened yet. It does not matter that we could, given enough information about the present state of the Universe (such as the positions and states of motion of all the particles in the Universe), *calculate* what will happen at all future times. This is no more than Newton's deterministic (clockwork) universe. The difference now is that distances, durations and the ordering of certain events will depend on the observer.

To view the whole of spacetime (the Universe at all times) as one 4D block requires a vantage point that is outside the Universe. This is the same as asking what the Universe looks like from the outside. There is no outside. So such a view is hypothetical.

These arguments have not stopped many physicists, mathematicians and philosophers from embracing the block universe idea, with its static time, wholeheartedly. The mathematician Hermann Weyl describes the block universe thus: "The objective world simply *is*, it does not *happen*. Only our consciousness . . . is as a process that is going forward in time". In fact, such a view was held long before relativity and is very close to the arguments put forth by the German philosopher Immanuel Kant in his *Critique of Pure Reason* of 1787.

I have always felt there to be an inconsistency in this argument. Weyl would have us believe that despite nothing ever *changing* in 4D spacetime, our consciousness still somehow moves *through* it, which is how we have the feeling of an ever-changing present moment. He claims that this feeling is illusory. But movement, however illusory, implies change, and change requires the passage of time. So if our consciousness experiences change then it must exist outside static spacetime. However mysterious consciousness is, I am not willing to attribute to it such status.

Gravitational times

Now that we have seen how time is described in special relativity I will discuss briefly how it behaves in general relativity. Einstein showed that gravity provides an alternative way of slowing time down to travelling at very high speeds. We have already seen how general relativity describes the way massive objects cause spacetime in their vicinity to curve (notice how I can finally talk about spacetime now rather than just space). Just as space gets stretched inside gravitational fields, so does time. Consider the effect on time near the event horizon of a black hole. An observer watching, from a safe distance, someone holding a clock while falling into the hole will see the clock run more slowly. This is why we see objects that fall into black holes appear as though they are frozen at the horizon. To us, time at the horizon is standing still. However, this is not simply an optical illusion. We have seen that the time dilation in special relativity is itself relative. Two observers moving at high speed relative to each other will each see the other's clock run slower. But in the case of the two observers near the black hole, the one falling in will see the clock of the distant observer running faster!

You are entitled to feel less than convinced by this discussion. After all, no one has come face to face with a black hole for such an effect to be tested. So how can we be sure that time would really slow down? The answer is that we can test it here on Earth. The gravitational field of the Earth is nowhere near as strong as a black hole's but we can still measure the tiny effect it has on time.

The dilation of time due to the Earth's gravity was confirmed in a famous experiment carried out by two Americans in 1960. Robert Pound and Glen Rebka made use of the recently discovered Mössbauer effect, which states that an atom of a particular type will emit light of a specific wavelength when pumped with energy. And because this wavelength is compatible with other similar atoms, they will readily absorb the light. If the wavelength is changed ever so slightly, say by a Doppler shift, then the other atoms will not be able to absorb it. Pound and Rebka placed some 'emitting' atoms of iron at the bottom of a 23 metre high tower and identical atoms at the top. They found that the light emitted by the atoms at the bottom was not absorbed by those at the top, and showed that the reason for this was that the wavelength of the light was redshifted. This 'gravitational redshift' is a direct result of the slowing down of time at the base of the tower. You see the top of the tower is further away from the Earth and gravity is therefore weaker there (not by much, of course, but enough to alter the wavelength of the light sufficiently.

To understand this redshift as a slowing down of time, consider what a wavelength actually means. You can think of the atoms of iron as clocks with each crest of a light wave they emit as a 'tick'. If we see longer wavelengths, it will be because more time has elapsed between successive ticks and we say the atomic clock is running slower. To measure how *much* time was slowing down, Pound and Rebka did something which I remember thinking, when I first learnt about it, was an absolute masterstroke. They made the atoms at the top of the tower move down with a specific speed towards the ones at the bottom. The moving atoms now saw the wavelength of the light travelling up to meet them slightly squashed due to the Doppler shift. This shortening of the wavelength could be adjusted, by controlling the speed of the downward-moving atoms, to restore the wavelength of the light to its correct value and the falling atoms were thus able to absorb the light.

In certain situations, the two time dilation effects (due to special and general relativity) can act against each other. Consider two atomic clocks, one on the ground and one in a satellite in orbit. Which will be running slower? To the clock on the ground, the high speed motion of the one in orbit should be making it run slower, while the fact that it is orbiting the Earth in zero gravity should be making it run faster. Which effect wins? The answer is that it depends on how high up the satellite is. Scientists need to know this sort of thing when analysing information sent down by navigational satellites which have their own atomic clocks. As an example, if a satellite is orbiting at an altitude that is more than the diameter of the Earth, it will be sufficiently far away for the gravitational time dilation to win. Its clock will be running faster than the clocks on Earth, which are slowed down by Earth's gravity, by a few millionths of a second each day (an unforgivable inaccuracy where atomic clocks are concerned).

So just remember, if your watch is running slow, hold it above your head! It will speed up now that it is feeling a weaker gravitational force. Of course you would never be able to measure such a tiny effect however long you hold you arm aloft.

7

TIME TRAVEL
PARADOXES

We have seen how an interval of time depends on your perspective in relativity. Less time will elapse for someone who accelerates to very high speeds or spends some time in a strong gravitational field. I have discussed how this provides us with a way of travelling into the future. Of course this cannot really be considered as cheating time. All we are doing is getting to the future more quickly. Think of it like those clever adverts on TV where someone is slowly munching through a chocolate bar while the rest of the world whizzes past at high speed. I have described how this sort of time travel is quite normal for subatomic particles, since they are the only objects capable of getting close to the speed of light. Thus the muons that are produced by cosmic rays are time travelling into the future (by a tiny fraction of a second) during their shortened journey through the Earth's atmosphere.

The problem with this sort of time travel is that it is one way. We may well be able to one day travel into the future, maybe even the distant future, but the only way to get back to our own time again would be by travelling *back* into the past. This is an altogether trickier problem. When scientists talk about time travel they tend to mean time travel into the past. Throughout this chapter, whenever time travel is mentioned it will refer to time travel into the past.

There are two ways of going back to the past. One is by going *backwards* through time, during which the hands on your watch

174

would be moving round anticlockwise. Of course you would not be aware this is happening. This would require faster-than-light speeds which are not accessible to us, and so is not the sort of time travel I intend to discuss here. The other way is by travelling forward in time (your local time runs forwards) but by moving along a warped path through spacetime that takes you back to your past (like looping the loop on a roller coaster). Such a loop is known in physics as a 'closed timelike curve' and has been the subject of intense theoretical research during the 1990s. What may come as a surprise to you is that it has been known for half a century that Einstein's equations of general relativity allow such closed timelike curves. The Austrian-born American mathematician Kurt Gödel showed in 1949 that such time travel into the past was theoretically possible.

So what is all the fuss about? Time travel to the future is easy and time travel to the past, while difficult, is not yet ruled out by theory. What are we waiting for? Why haven't we built a time machine yet? The reason is that not only would it be exceedingly difficult to create a closed timelike curve in spacetime, but that we are not sure whether it is even possible in theory. As things stand, general relativity tells us that we cannot rule out time travel, but many physicists are hoping that a better understanding of the mathematics will eventually lead to the conclusion that it is totally forbidden. And the reason physicists feel so strongly about this is that time travel leads to a number of strange paradoxes. In this chapter I will take a look at some of these time travel paradoxes and see if there is any way out.

As a scientist, I find it hard to just sit and watch a science fiction film that involves time travel. Instead of taking it with the requisite pinch of salt as I am supposed to, and just enjoying the (usually) daft story line, I tend to pick holes in the logic. I'll say something like: hang on a minute; if he just went back in time and did such and such then surely he has meddled with history and... well, you probably know what I mean.

It's quite sad really.

Most of these films are rather silly and I should just 'go with the flow' and appreciate the millions of dollars spent on the special effects. If you've ever seen *Star Trek IV: The Voyage Home* you'll know what I mean. In that film, the best of the original Star Trek movies, Captain Kirk and his crew travel back in time to the twentieth century. It contains some amusing moments, like when Scotty tries to talk to a computer, is told that he must use the mouse and so picks it up and talks into it! Well it's my favourite bit.

The Terminator paradox

I want to use a variation on a story line in a particular film to illustrate a time travel paradox. That film is the *Terminator*, in which Arnold Schwartznegger is an indestructible android sent back in time by the robots that rule the world in a violent future. John Connor is the name of the rebel leader who is fighting for the humans' cause against the robots, and Arnie is supposed to kill John's mother before she has given birth to him. You see if John had never been born in the first place then the rebels would be easily defeated. So, by bumping off his mum, they get rid of him too.

Of course not only does Arnie fail, but the hero of the film, who is sent back in time to protect John's mum, ends up falling in love with her, gets her pregnant and she gives birth to... John Connor. So this guy, who is the same age as John in his own (future) time, is actually his father. He was sent back to make sure that John is born, and ends up being the *reason* he is.

The question is whether it would have been possible for Arnie to have altered the course of events so that the future could turn out differently? What if he had succeeded in killing John's mum? Even though the story line in the film may sound rather silly if you are not a sci-fi fan, it is nevertheless consistent. No paradox arises because Arnie fails. Come to think of it the film is not at all bad and the special effects are brilliant. (Yes, I know they are even better in *Terminator II*.)

I wish here to retell the story in order to highlight the most famous time travel paradox, known as the grandfather paradox.

Stated in its original form, the paradox arises when you go back in time and murder your grandfather before he meets your grandmother. So your mother was never born and neither were you. And if you were never born your grandfather *could not* have been murdered by you, and so you *would* have been born, so he *was* murdered, etc. This is the paradox. The argument keeps going round in a self-contradicting circle.

Let us rewrite the script for the *Terminator*. Suppose that the robots from the future decide that, instead of sending their muscle-bound android back in time to do the biz on Mum, they will capture John Connor himself, and persuade him by whatever foul means to go back in time himself and murder his mother.

What happens if he succeeds? I mean, if she is killed before she gives birth to John then he never existed. Does he simply fade away as she slumps to the ground? And if he never existed then who killed his mother? It couldn't have been John; he was never born!

There are several ways out of this one. All have been aired in many science fiction stories in one form or another. I will consider three scenarios:

1. *As John shoots his mother he 'pops' out of existence.* This simply won't do. His mother will have a bullet imbedded in her heart (doesn't mess around does he, our John) that must have been fired from a gun. Someone pulled the trigger. You can't say that John existed before he shot her because now that he has altered history he was never born in the first place. The past will have evolved differently (Johnlessly) and there will have been no need for the robots to send someone—especially someone who doesn't exist—back in time to kill Mum. This explanation implies that there are now two versions of history: one in which John was born and one in which he was not, which cannot be right.

2. *John cannot murder his mother because he is there to try.* In other words, the fact that he exists means that any attempt he makes *must* fail. This is certainly better than the first option since it ensures that there is only one version of history. However it still leads to a problem, as we shall see.

3. *When John goes back in time he slides into a parallel universe; one in which he is allowed to alter the course of history.* Thus, even though he cannot change his own past he can change the past in a neighbouring, yet almost identical, universe. So when he kills his mother he will never have been born in *that* universe but his mother would have continued to live in his own. This type of explanation has, until recently, only been popular with science fiction writers. But, believe it or not, it is now being taken seriously by some physicists who would like parallel universes to exist for quite different reasons. I will come back to this later on and show that it is the only viable way out.

The simplest and, many would say most reasonable, of the above options is the second one. Let us assume that there are no parallel universes (since there is no evidence for them and, in the absence of any time machines, no way we can check). There is also only one version of history. We cannot go back and change its course since we already *remember* events from our past. Basically, what has happened has happened.

This is not the same as saying that we are unable to go back in time and meddle with the past. It's just that if we do, we must have caused things to turn out the way they have. So a time traveller can never go back and stop J F Kennedy from being murdered, but could himself have been the murderer.

This way of explaining how we could go back and participate in our own past is exploited wonderfully in the *Back to the Future* films. There, certain events that happen are not explained at the time. Only later do we learn that they were caused by characters travelling back from the future. So we see certain scenes twice: first from the point of view of the characters living in their own time and then from the point of view of the time travellers (usually older versions of the same characters).

Back to my version of the *Terminator*. John may well try to shoot his mother but clearly something has to happen to stop him. This may be due to any one of a number of reasons. Maybe he comes to his senses in the nick of time. Maybe the gun wasn't loaded, or the trigger gets stuck. Maybe he is just a lousy shot. It doesn't really matter why he fails, simply that he must fail. His mother has to survive for him to be there in the first place. I will refer to this sort of puzzle as a 'no choice' paradox, since it suggests that time travellers do not have the freedom of will to do certain things which will alter the course of history in such a way as to make it impossible for them to have travelled back in the first place. Once you start thinking about this you realize there is a real problem. Does this mean that if John were to try again and again he will always be doomed to failure? We can imagine that the robots return John back to their time, give him a double dose of 'kill your mother' serum, intensive shooting lessons and send him back to the past with a well oiled, definitely loaded, foolproof, sawn-off shotgun. He will still fail. The laws of physics are not required to explain why he will always fail. All that matters is that paradoxes are avoided.

Theoretical physicists have devised a thought experiment to see what would happen in a real situation if something were to travel back in time and meet itself. What would the mathematics predict? To make the model sufficiently simple, they came up with the billiard table time machine. The idea is that a ball enters one pocket of a billiard table and emerges from an adjacent pocket in the past. It can therefore collide with itself before it went in. In this model, all paradoxes can be easily avoided if we allow only those situations that do not lead to a paradox, called 'consistent solutions'. Thus a ball can go back in time, pop out of another pocket and deflect the earlier version of itself into the hole, enabling it to travel back in time in the first place. But the situation in which the ball emerges from the pocket and collides with its earlier self such that it causes it to miss the pocket it would have rolled into, would not be allowed mathematically since it leads to a paradox. This is all very neat and means that supporters of time travel can pat themselves on the back for proving mathematically that paradoxes *can* be avoided if we are careful. The problem they try to pretend does not exist is avoided because billiard balls do not have free will. They have not exorcized the no choice paradox.

The rule is therefore that the past has happened and we are allowed only the one version of it. We may do whatever we like when we travel back in time as long as we remember that however much we meddle with history we will always cause it to turn out the way it has. Even in the original story line, Arnie can never succeed because he comes from a future where John is alive, and therefore should not even bother trying so hard. But then I suppose the film would not be anywhere near as enjoyable.

Trying to save the dinosaurs

Remember I mentioned that we can still 'meddle' with the past? Let me put it another way: we are allowed to go back to the past and cause things to turn out the way they have *because* of our meddling. I will illustrate this with another example.

Scientists are mostly in agreement now that a large meteorite hit the Earth over sixty million years ago and that the fall-out from the impact caused a dramatic climatic change that killed the dinosaurs. Some mammals did, however, survive and some of these evolved into apes and then humans millions of years later. In fact, we can probably say that, had the dinosaurs not been wiped out then mammals, and hence humans, would not have been allowed to evolve. Put another way, it is thanks to that meteorite that we are here at all.

Suppose a palaeontologist gets hold of a time machine and a thermonuclear missile (unlikely I know, but bear with me). He travels back 65 million years into the past intent on destroying the meteorite before it hits the Earth. But if he saves the dinosaurs from extinction then not only will he 'pop' out of existence but so will the rest of the human race. This is a pretty extreme form of the grandfather paradox.

I should come clean at this point and say that if we do ever build a time machine then the laws of physics definitely do not allow us to travel back to a time before the time machine was built. This is because constructing a time machine involves linking different times together within spacetime. So the earliest time that is linked in this way will be the moment of the time machine's creation. All times before this would have been lost forever and no longer 'available'. This rules out any possibility of us ever being able to go back to prehistoric times—unless we stumble upon a cosmic time machine somewhere in space that has been around for a long time.

For the sake of argument let us say that the mad palaeontologist does manage to go back to a few hours before the impact of the meteorite and points his missile at it. He looks up and sees to his dismay that it is much larger than he expected. In fact, if it collides with Earth, as it seems rather intent on doing, it will wipe out all life, not just the dinosaurs.

"Well," he thinks to himself, "I might as well do what I can." He fires the missile and scores a direct hit. The meteorite is destroyed.

But... it seems a small fragment survives and is still heading towards Earth. Not having any more missiles there is nothing he can do now but watch the fireball as it streaks through the atmosphere, hits the Earth and, guess what? causes the extinction of the dinosaurs!

So you see, not only was he unable to change history, he was actually the cause of it. Had he *not* gone back and destroyed the meteorite he would never have existed. I have turned the argument upside-down and the paradox seems to have disappeared. But unlike the last story in which John must keep missing however hard he tries to shoot his mother, now our palaeontologist cannot miss, for if he does he will never have existed. He cannot decide at the last minute not to fire the missile, nor can anything deflect it from its course. It is easy to argue away any hint of the no choice paradox in this case by claiming that, unlike the previous example where we require something to cause him to fail, here the time traveller is unaware of the enormity of his task and of his lack of choice in the matter.

Mona Lisa's sister

The year is 1504, the place Florence, Italy, and Leonardo da Vinci has just completed his greatest masterpiece, the *Mona Lisa*. He decides to take a break from art and, being a bit of an all-round polymath-cum-jack of all trades, he decides to devote some

181

time to his other great love, inventing. After many weeks of contemplation, and long nights of ever more detailed sketches, he finishes the plans for his cleverest contraption yet: a time machine. After many more weeks locked away in his workshop he finally completes it late one evening, and goes to bed that night feeling rather uneasy as he ponders how to test it out.

The next morning his unease is vindicated when he finds to his astonishment a painting in the time machine. It is a portrait of a woman who bears a close resemblance to his Mona Lisa: the same facial features and long dark hair, but without the enigmatic smile. He immediately recognizes the woman as Mona Lisa's ugly sister, Mona Lot, who had been pestering him to paint her and whom he had been avoiding for some time. This painting is clearly his own work (it even has his signature) and, since he knows he has not painted it, he deduces that it must have been sent back in time from a future Leonardo. Of course he is thrilled by this. He decides not to tell anyone but to keep the new painting under wraps while he figures out what to do.

As the days go by Leonardo becomes more and more worried about possible time travel paradoxes. He knows he must send it back because that is how it came into his possession. On the other hand, if and when he does he loses it forever. It will be stuck in a time loop.

There are two different paradoxes here. The first is a no choice paradox since Leonardo knows he must at some time send the *Mona Lot* back in his time machine. Let us suppose there was a note with the painting specifying the time and date it was sent. He knows as he approaches that day that whatever he may try to do to avoid sending it back he will fail. What if he tries to cheat time by destroying the painting? This a very severe form of the no choice paradox. We have no problem with our inability to alter the past and use phrases such as 'it's no use crying over spilt milk', and 'what's done is done'. Here though, the future is linked directly to the fixed past and is itself therefore fixed. So what is it that will stop Leonardo from destroying the Mona Lot? What unknown force will protect it from being burnt, slashed or thrown into the river Arno?

There is another kind of paradox which to many is even more disturbing than the no choice paradox and which I will refer to as the 'something-from-nothing' paradox. It arises even if Leonardo does send the painting back at the allotted time, and then destroys the time machine before it can give him any more headaches. We are still left with a puzzle, namely *who created the Mona Lot?* Leonardo may now feel that he has weathered the storm and that, whatever strange obstacles there may have been to thwart his possible attempts to force a paradox, the whole episode is now thankfully in his past. But there is no getting away from the fact that, for a while, there existed a Leonardo da Vinci masterpiece which *at no time did Leonardo da Vinci actually paint!* He found it in his time machine, kept it for a while before putting it back into the time machine and sending it back to himself. But where did it *originally* come from? Apparently nowhere. It was caught in a time loop and Leonardo never painted it! No amount of arguing about ensuring logical consistency can lay the something-from-nothing paradox to rest.

No way out?

You can see why so many physicists do not accept that time travel to the past will ever be possible. There is yet another paradox that I have not mentioned which is to do with using a time machine to create multiple copies of yourself, thus violating sacred laws of nature like conservation of mass and energy. For instance, you could travel back to five minutes ago and meet yourself. Are you then able to both enter the time machine and go back five minutes earlier to meet a third you and so on? This is really just another form of the no choice paradox. As there is only one version of the past and you know before you first set off that no copy of you arrived five minutes ago from the future, you cannot be free to join yourself. You cannot go back to meet yourself because you have no memory of meeting yourself.

It seems that I have provided more than enough nails in the time travel coffin and you may be wondering whether it is worth sticking with the rest of the book in which I explain how a time machine may be built. But don't despair just yet. It is clear that there have to be certain ground rules about which times we are allowed to go to and what we are allowed to do in order that paradoxes cannot arise. Many die-hard time travel fans are not too worried about the no choice paradox. They agree that we must sacrifice free will if time travel to the past is to be possible. For them, we do not have free will anyway, we just think we do. Since we live in a deterministic universe where everything is pre-ordained anyway, we do not need to appeal to anything new. As long as everything is logically consistent there is no problem. Thus, you can go back one hour to meet your (slightly) younger self *if* you remember meeting your (slightly) older time travelling self an hour ago. If you don't remember it then you will not be able to travel back. Not even the something-from-nothing paradox deters such ardent supporters. "So what," they say, "the Mona Lot painting being caught in a time loop does not give rise to any logical inconsistency."

For me though, this is a much more serious abandonment of free will. At least in a deterministic universe we are under the illusion that we are making our own free choices and decisions. In time travel, we are not allowed the luxury of this feeling. Our free will is wrenched from us in a way that is far from clear.

If you are determined to allow for the possibility of time travel into the past then there is another price you must pay. You have to take the block universe description of spacetime seriously. Past present and future must all coexist. The reason for this is simple: If you go back into the past (which we reluctantly admit might exist since at least we know it did exist) then for those people you meet when you travel back in time (even a younger version of you) that time is their 'now', their present moment. You have arrived from their future. They would have to admit that the future is just as real as the present. We cannot even claim that our present is the true 'now' and that they just think they live in the present, since we can similarly imagine time travellers from our future visiting us in our present. If they do, then our future, and indeed all times, must exist together. This is precisely what the block universe model tells us.

Parallel universes

If you are not prepared to sweep time travel paradoxes under the carpet, there is an alternative possibility in which neither the no choice paradox nor the something-from-nothing paradox need ever appear. The price to pay may be too high for you to accept though, despite the inevitable desensitization you must be undergoing towards some of the absurdities you have already met in this book. What I am about to describe is going to sound crazier than anything else you have met so far, yet it is based on a highly respectable, if unconventional, interpretation of the weird results of quantum mechanics.

I have already briefly mentioned at the beginning of Chapter 4, when describing the nature of light, that quantum mechanics ranks even higher than relativity in terms of its importance as a scientific discovery which has affected our everyday lives. The problem is that nobody really understands what it is telling us about the world of the very small which it describes so accurately. This will of course sound rather strange. How can a theory which we do not understand be so successful? The answer is that it predicts the behaviour of the very building blocks of matter—not just the atoms, but the particles that make up the atoms (the electrons, protons, neutrons) as well as light photons and every other subatomic particle you care to name (and there are many)—with incredible accuracy. Quantum mechanics has led us to our current very precise understanding of how these 'quantum' particles interact with each other and connect up to form the world around us. Yet, at the same time quantum mechanics forces upon us a view of the subatomic world which goes totally against our common sense. Physicists have now had three quarters of a century to come to terms with this and have come up with a number of possible interpretations of what must be going on down at the quantum level. But there is still no overall consensus as to which is the correct interpretation, if indeed there is only one.

The interpretation that has held sway for most of the twentieth century is known as the Copenhagen interpretation (as it originated at the institute in Copenhagen where one of the founding fathers of quantum mechanics, Niels Bohr, worked).

Its supporters take a very pragmatic view of the whole issue by claiming that we should not worry about trying to understand what is happening down at the scale of atoms which is so far removed from our everyday world. For instance, we have no right to expect a photon of light to behave in a way that makes sense to us. If light appears one minute to have the properties of a stream of particles and the next those of a wave then so be it. All that matters, say the Copenhagenists, is that quantum mechanics works. The mathematics agrees beautifully with what we see around us in the real world, so why beat ourselves up over it?

Such a view has, until the last ten to twenty years, been the majority one. Most practising physicists have been happy (well, maybe happy is not the right word) to use the tools of quantum mechanics—abstract symbols and mathematical techniques rather than spanners and screwdrivers you understand. They have been prepared to leave the pondering and musings about the deep meaning of it all to the philosophers.

The differences between the interpretations are to do with how we describe what is 'happening' to a quantum particle, such as an electron, when we leave it alone to do whatever electrons like doing when left alone. If we measure a certain property of an electron, such as its position, speed or energy at a particular moment, then quantum mechanics will tell us what we are likely to find. However it tells us nothing about what the electron is doing when it is not being observed. This would not be a problem if we could trust electrons (and all other quantum particles) to behave sensibly, but they don't. They will disappear from the place they were last seen and spontaneously reappear somewhere else that should, by rights, be inaccessible to them. They exist in two places at once, they tunnel through impenetrable barriers, travel in two different directions at once and even have several conflicting properties simultaneously. But the moment you look to see what is going on, the electron will suddenly start behaving itself again and nothing will look out of sorts. However, the unavoidable conclusions we have to draw from the results of our observations is that the electron was most definitely doing something very strange indeed when we weren't looking. In fact all quantum particles behave in ways which would be quite impossible if they obeyed the same rules as everyday objects we are used to.

Since quantum mechanics only ever tells us what to expect from the results of our observations we must appeal to something else if we insist on trying to understand what is going on when we are not looking. This is what I mean by different interpretations. I have so far only mentioned the Copenhagen interpretation, which for many has been considered the standard one. It is the one which almost all textbooks on quantum mechanics use and which is taught to all physics students. But there is a growing consensus that its time is up. Its supporters may still argue that it is purely a matter of philosophical taste which interpretation we choose but the truth is that a growing number of physicists (not quite a majority yet) are looking for something deeper.

One of the alternative explanations to the Copenhagen view, which is of particular interest to time travel fans, is known as the many-worlds interpretation. According to this view, as soon as a quantum particle, anywhere in the Universe, is faced with a choice of two or more options, the whole Universe splits into a number of parallel universes equal to the number of options available to the particle. There are, according to this view, an infinite number of universes which differ from our own to a greater or lesser degree depending on how long ago they split off from ours, and each universe is just as real as our own. In many of these universes there exist carbon copies of you. In some you are a billionaire business tycoon, in others you are living rough on the streets. In many others you have turned out very much like you have in our Universe apart from some minor details. For many years, this sort of thing has been the stuff of fiction, and for many physicists it will remain that way. There is no experimental evidence whatsoever that parallel universes exist since we cannot make contact with any of the others, but the truth is that it has its good points as a scientific theory and explains away much of the strange behaviour of the quantum world. But at what price? Physicist Paul Davies has remarked that the many-worlds interpretation is cheap on assumptions (a point in its favour) but expensive on universes.

If you are meeting this idea for the first time it might seem impossible that there is even any room for all these other universes.

187

After all, it may well be that our own universe is itself infinite. Where can all the others be? The way to visualize this is to think of our Universe as an infinitely extended flat sheet by throwing away two of its dimensions (remember we have to go from four-dimensional spacetime down to a two-dimensional sheet). Now the other parallel universes can be thought of as stacked above and below our own. There is room enough for all with enough dimensions.

Apart from the sheer extravagance of requiring an infinite number of universes, some physicists claim that the many-worlds interpretation also does away with free will. Here is how it works: Whenever you are faced with any kind of choice, say touching the tip of your nose and not touching the tip of your nose, and you choose (you think freely) not to touch it, what will have happened is that the Universe split into two and there will now be a parallel universe in which you did touch your nose. You are conscious of having taken one of the available pathways, but there will be another version of you in a parallel universe who is conscious of having made the alternative choice.

The many-worlds interpretation of quantum mechanics was proposed by an American physicist by the name of Hugh Everett III in the 1950s and, despite not catching on at the time, has recently been favoured by a growing number of cosmologists who feel it is the only viable interpretation when applying quantum mechanics to describe the whole Universe.

At the same time, the idea of parallel universes has been exploited by science fiction writers who have recognized that it rescues them from time travel paradoxes. More recently, the Oxford physicist David Deutsch has developed his own version of the theory and points out that if we wish to take the possibility of time travel into the past seriously then we are forced to take the many-worlds interpretation seriously. I should explain that Deutsch is a strong supporter of the many-worlds idea, which he refers to as the multiverse interpretation (the term 'multiverse' implies the multitude of all universes, the totality of reality). I am not convinced by this view, but I cannot rule it out. For what it's worth, my favourite interpretation of quantum mechanics is one due to the physicist David Bohm which requires just the one universe thank you very much.

So how does the many-worlds view cope with the paradoxes of time travel? As I mentioned in option three when attempting to explain the possible alternatives available to John Connor in the Terminator paradox, a time traveller will not travel into the past of his or her own universe—not surprising given that there is an infinite number of pasts to choose from—but into that of a parallel universe.

According to Deutsch, who takes the block universe idea literally in his book *The Fabric of Reality*, the Universe does not divide up into multiple copies of itself at the moment we are faced with a choice. Instead, there are already an infinite number of parallel universes out there. At the moment of choice we are just following one particular pathway, like a train going through a complicated junction. This means that the future is open since there are many options available to us, but so is the past. Our own spacetime is just one of an infinite number of pasts *and* futures. Travelling into the past in Deutsch's multiverse is no different to the way we would normally get carried along into the future. We simply follow a time loop into one possible past.

Deutsch's approach is just one of a number of versions of the many-worlds interpretation. I will re-examine our time travel paradoxes within the more conventional version of the many-worlds theory—if you can ever call an infinity of parallel universes conventional. The no choice paradox no longer applies since we are free to alter the past as we wish since it will not be our past. Events in the parallel universe we have travelled to need not turn out the way they did in our own universe. John Connor can now kill his mother and stop himself from being born *in that universe* while the mother back in his own universe survives. Of course if he does fail to kill her then he will be born. There will now be one universe (soon to proliferate due to all the other quantum choices that are going on) in which John Connor grew up, hopped into a time machine and disappeared forever. There will also be another universe in which there is either one or two 'John Connors', depending on whether he succeeds in killing his mother or not.

If he does not kill her the two Johns will live side by side but separated in age by however far back he has travelled in time. One thing to remember though is that the chances of John finding his way back to his own universe, or even one very much like it, are very small indeed. There are simply too many to choose from.

The something-from-nothing paradox can also be explained away. The Mona Lot painting was of course sent back from the future of a parallel universe and was painted by the Leonardo in that universe. Leonardo does not even need to send the painting back to the past at the allotted time. He can keep it. After all, even if he does send it back it would be a third Leonardo who will find it in his time machine.

Even the problem of creating multiple copies of ourselves by looping repeatedly around in time is resolved. If there ends up being 100 copies of you in one universe this just means that there are 99 other universes from which you have disappeared. Conservation of mass and energy no longer applies to each universe separately but to all universes taken together.

One of the advantages of the original version of the many-worlds interpretation, in which the Universe splits only when you are faced with a choice, is that it gives us an arrow of time which points in the direction of increasing number of universes. There are always more universes in the future than the past. But surely, you might think, isn't this rule violated when we allow for time travel into the past? If I travel back today into the yesterday of a parallel universe I cause it to start splitting according to the choices I can make once I am there. But how can that parallel universe have started splitting yesterday before I even made the decision to travel back? Is this another form of the no choice paradox? It seems as though the parallel universe I am about to travel to must know in advance that I will be arriving and making certain choices, thus forcing me to travel back to that universe, and making those choices.

Again, the many-worlds interpretation offers a neat way out. General relativity allows a way of connecting up our Universe with a parallel one which may not necessarily involve time travel into the past, as we shall see in the next chapter when I introduce wormholes. A time machine is of course one way of making this connection. The mistake in the previous paragraph is to think that this connection is made the moment we travel back in time. It is not. It is made the moment the time machine is created (or switched on) allowing for the *possibility* of time travel to any universe which splits off from ours subsequent to the moment the time machine is switched on. At the instant of switching on there will be universes in which versions of us start arriving. This is because, even if we decide not to use the time machine but to destroy it instead, it is too late. Having been faced with this choice, our Universe splits and there is another universe in which we did not destroy the time machine but used it instead. In some universes we travelled back to the very earliest moment possible (just after it was switched on). In others, we travelled back to a later time. Even while you are still strapping yourself into the time machine the Universe is splitting, due to all the other choices it is making everywhere else in space, and there will therefore be an infinite number of '*you*'s travelling back in time! I think I'll stop typing now and go and lie down for a while.

Where are all the time travellers?

I hope I have given you an idea of the problems we must face if we insist on the possibility of time travel into the past. The laws of physics as we understand them do not rule it out, so where is the flaw in the argument? You might feel that having to either live with the time travel paradoxes or to accept the notion that there is an infinite number of parallel universes is just too much. But even physicists have failed to come up with a more convincing argument to rule it out. One that you may have come across is to ask where all the time travellers from the future are? If future generations ever succeed in building a time machine then surely there will be many who would wish to visit the twentieth century and we should see these visitors among us today. I will therefore list five possible reasons why we would not expect to see any time travellers:

BLACK HOLES, WORMHOLES & TIME MACHINES

1. Time travel to the past is forbidden by some as yet undiscovered laws of physics.
2. A time machine can only take you as far back as the moment it was switched on and no earlier. So if we figure out how to build a time machine in the twenty third century we will not be able to visit the twentieth century. The only way that would be possible is if we come across a naturally occurring time machine that has been around for long enough, such as a black hole or a wormhole. Maybe there are none to be found in our neck of the Universe.
3. Naturally occurring time machines are found and people do use them to travel back to the twentieth century, but it turns out that the many-worlds theory is the correct version of reality. Our Universe is just not one of the lucky few which visitors have visited.
4. Expecting to see time travellers among us presupposes that they would, in fact, want to visit this century. Maybe for them there will be much nicer and safer periods to visit.
5. Time travellers from the future are among us but keep a low profile!

Much as I would like to think that time travel is possible, I am afraid I would probably put my money on the first point. The reason for this is really quite straightforward and I have mentioned it before. For time travel to the past to be possible, the future—our future—has to be already out there. I find this hard to accept.

Don't despair. My advice to you, if you do not want to give up on time travel, is to take comfort in the fact that there remain loopholes in the laws of physics which allow it. As long as time travel is not categorically forbidden we shall continue along our journey.

TIME MACHINES

8
WORMHOLES

"Look fellas," she said, "I'm no expert in General Relativity. But didn't we see black holes? Didn't we fall into them? Didn't we emerge out of them? Isn't a gram of observation worth a ton of theory?"

"I know, I know," Vagay said in mild agony. "It has to be something else. Our understanding of physics can't be that far off. Can it?"

He addressed this last question, a little plaintively, to Eda, who only replied, "A naturally occurring black hole can't be a tunnel; they have impassable singularities at their centers."

Carl Sagan, *Contact*

Our journey has taken us from the beginning of time to the very edge of the Universe. The legacy left us by Albert Einstein describes a reality far more wonderful and mysterious than anything we could have dreamt up. Time warps, black holes, parallel universes, a past and a future that coexist with the present, none of these are the stuff of science fiction. Nor are they the results of the wilder speculations of a nutty minority on the fringes of the scientific establishment. All these exotica are the results of years of slow progress, some of which are now regarded as facts. For instance, the slowing down of time due to gravity is not 'just a theory which may turn out to be wrong tomorrow when something better comes along', but is shown to be true on a regular basis in scientific laboratories. Other ideas, while possibilities, may not stand the test of time or the continued close scrutiny of scientists. Sometimes a theory is shown to be just plain wrong if its predictions conflict with the results of an experiment, or it may be replaced with a better theory which explains more phenomena and gives us a deeper understanding of nature.

We are now reasonably confident that black holes exist. This is despite the fact that we have never come face to face with one. The evidence for them is so convincing that we cannot find an alternative explanation. Not only are black holes an inevitable consequence of the theory of general relativity but we see their unmistakable signature through our telescopes.

Wormholes are a completely different matter. They are also allowed by the equations of general relativity which give a description of them as theoretical entities. But, unlike black holes, wormholes remain theoretical curiosities with not a shred of evidence from astronomy for their existence in the real Universe. I am sorry to be spoiling all the fun by pouring cold water on objects that I have not even discussed yet. Maybe this is just my defence mechanism against accusations from other physicists that I am sailing close to the edge between science fact and science fiction. Thus in order to justify this chapter to those cynics who are more conservative in their views I will quote a short passage from the beginning of Matt Visser's book *Lorentzian Wormholes: from Einstein to Hawking* with my additions in square brackets:

> "Even though wormhole physics is speculative, the fundamental underlying physical theories, those of general relativity and quantum [mechanics], are both well tested and generally accepted. [Even] if we succeed in painting ourselves into a corner surrounded by disastrous inconsistencies and imponderables, the hope is that the type of disaster encountered will be interesting and informative."

Thus it may well be that wormholes do not exist, but at the very least their study might help us to understand a little better the way our Universe works. Oh, and in case you are wondering, they have nothing whatsoever to do with worms.

A bridge to another world

The idea of wormholes dates almost as far back as general relativity itself. Remember from Chapter 4 that Karl Schwarzschild was the first to realize that Einstein's equations of general relativity predicted the existence of black holes. More specifically, his black hole contained a singularity at its centre; a point of infinite density where time itself came to an end. At the singularity, all the known laws of physics break down. This troubled Einstein. He didn't like these holes in spacetime and it was not enough for him that they were shielded from the outside world by event horizons. For him it was not simply a case of 'out of sight, out of mind'.

In 1935, Einstein published a paper with his collaborator Nathen Rosen in which they attempted to prove that Schwarzschild singularities did not exist. By using a mathematical trick known as a co-ordinate transformation, they were able to rewrite Schwarzschild's mathematical solution so that it did not contain a point where space and time stopped. The alternative, however, was just as strange. They showed that the singularity became a bridge connecting our Universe with... a parallel universe! This is not the sort of parallel universe that would have split off from ours as a result of quantum mechanics as I described in the last chapter. This link between the two universes became known as the Einstein–Rosen bridge. It was, for Einstein, a purely theoretical exercise in geometry in which two spacetimes would be joined together. He did not believe that such a bridge really existed, any more than he believed singularities really existed. It was just an oddity of the mathematics of general relativity.

Such bridges between different worlds were not new even then. The mathematicians of the nineteenth century were very keen on curved space and higher dimensions. In fact, exactly half a century before Einstein published his work on general relativity, an English mathematician by the name of Charles Dodgson wrote a children's book on the subject of higher dimensional geometry and parallel universes. Under the pen name of Lewis Carroll, he wrote *Alice's Adventures in Wonderland* in 1865. We are all familiar with the bit when Alice chases the white rabbit down an Einstein–Rosen bridge into another universe. I believe it is referred to in the book as a rabbit hole, but it means the same thing. The reason such strange things could happen in Wonderland was because the laws of physics were different in that universe. Of course

Dodgson was unaware what sort of mechanism could cause such a tunnel to join our world with another. Remember this was before relativity, quantum mechanics and modern cosmology. The story was based solely on geometrical ideas about how space could be curved and how two spaces could link together in some higher dimensional hyperspace. What Einstein showed fifty years later was that such curvature of space occurs wherever there is a strong enough concentration of mass (or energy since it is equivalent to mass). His theory of gravity (general relativity that is) provided the physical basis for such tunnels to other worlds even though they were no more likely to exist in reality.

In one of Dodgson's very last works, *Sylvie and Bruno*, which was published in 1890, we nevertheless find that he (and presumably therefore other mathematicians at that time) was also thinking about shortcuts within the same universe. In that story, Fairyland and Outland are a thousand miles apart but are linked by a 'Royal Road' which could take you from one to the other almost instantly. He also describes time travel, changing clock rates and the reversal of time.

Back to the 1930s and the reason no one was too excited about the Einstein–Rosen bridge was that, unlike the rabbit hole in *Alice's Adventures in Wonderland*, it could never be used as a practical means of getting to another universe. One way to think about how an Einstein–Rosen bridge could form would be to imagine a singularity in our Universe attaching itself to a singularity in the parallel universe. So could this be what would happen if we were to fall into a black hole? Think of black holes as a bit like the afterlife. Nobody really knows what awaits them when they die and, in the same way, we cannot be sure what will happen to us when we jump into a black hole until we actually do. Even then we are unable to relay the news back to those waiting outside the event horizon. As a scientist I would like to think that we know a little bit more about black holes than the afterlife since at least the former obey mathematical equations!

So what is wrong with the Einstein–Rosen bridge as a means of getting across to another universe? Well, to begin with there is the event horizon. Once you jump into a black hole you cannot come back out again. Of course, in order to come out the other side, the black hole you jump into would need to be hooked up to a white hole. Remember that this is the opposite of a black hole from which matter will emerge rather than fall into. White holes must therefore be surrounded by the opposite of an event horizon, something known as an antihorizon, which would allow one-way traffic out and never in. Unfortunately, antihorizons are very unstable and get converted to normal horizons in a matter of seconds after forming. So, having passed through the event horizon of the original black hole you would find that there is a second event horizon blocking your exit at the other end. Imagine a prisoner in a locked cell who discovers a tunnel under his bed. It leads underground for a few metres only to come out the other end inside an adjacent locked cell.

The major problem with the Einstein–Rosen bridge is that the whole thing is highly unstable. The connection would only survive for a fraction of a second before pinching off. In fact, so short is the lifetime of the bridge that not even light travels fast enough to get through. So if you *were* ever to jump into a black hole in the hope of getting across, you would always get caught in the singularity, and having one's body squeezed down to a size much smaller than an atom is never very desirable.

All this is assuming you weren't ripped apart by the tidal forces of gravity before you reached the singularity. The black hole would have to be a supermassive one for you to even survive going through the horizon. All in all, Einstein–Rosen bridges could never become a means for visiting a neighbouring universe, and therefore remained just a theoretical curiosity for many years.

Alice through the looking glass

There have been a number of landmarks in the history of wormhole physics. After the work of Einstein and Rosen in the mid-1930s nothing much happened until John Wheeler, one of the greatest physicists of the twentieth century (and the man who coined the name black hole) published a paper in 1955. In it he showed for the first time that a tunnel in spacetime need not necessarily join our Universe with a parallel one, but could bend round to join two different regions of our Universe together (like the handle on a coffee mug). It would be a tunnel that rose out of normal spacetime providing an alternative route between its two 'mouths' through a higher dimension. Two years later he introduced the word 'wormhole' into physics jargon in a landmark paper on what he called 'geometrodynamics' which means the study of how the geometry, or shape, of space changes and evolves. Of course his work was still purely theoretical. Its aim was to understand what shapes spacetime could be twisted into and had nothing to do with the use of wormholes for humans to travel through. In fact, the wormholes that Wheeler was interested in were extremely tiny ones. He was studying the structure of spacetime on the minutest possible scale where quantum mechanics tells us that everything becomes fuzzy and uncertain. Down at this level, even spacetime becomes frothy and foamy and all manner of strange shapes and structures, including miniature wormholes, can form at random. I will refer to these as quantum wormholes and we will meet them again a little later on.

The next important event was in 1963 when New Zealand mathematician Roy Kerr discovered that Einstein's equations predicted the existence of a completely new kind of black hole: a spinning one, although he did not realize this at first. Only later was it realized that Kerr's solution applied to any spinning star that collapsed to a black hole and that, since all stars are spinning on their axes at various rates, Kerr's black holes were more general and more realistic than Schwarzschild's non-spinning ones. What is more, a black hole would spin much more rapidly than the original star it formed from because it is so much more compact. (Remember I drew the analogy with the spinning ice skater when I described such black holes in Chapter 4.) What was so interesting about the results of Kerr's calculations was the nature of the singularity at the centre of such a black hole. It would no longer be a zero-sized point like those at the centres of Schwarzschild black holes but would be ring-shaped instead. The perimeter of the ring is where all the matter is, and has almost zero thickness and hence nearly infinite density. The middle of the ring is just empty space. Such a ring singularity could, depending on its mass and spin, have a large enough diameter for humans and even their spaceships to travel through[1].

Oxford astrophysicist John Miller has pointed out that, while Kerr's solution uniquely represents the properties of spacetime *outside* any stationary rotating black hole, there is as yet no indication whatsoever that it correctly describes what goes on *inside* the horizon, including everything about the ring singularity. It is thus just one possible picture of what the inside of a black hole might look like. Miller suggests that such descriptions should come with a government health warning.

With this in mind, I will go ahead and describe what a Kerr black hole might be like. To begin with, the ring singularity differs in other ways from Schwarzschild's point singularity. For instance, a ring singularity has a second, inner, horizon, called the Cauchy horizon, which surrounds the singularity. Of course once you pass through the outer event horizon there is no way back for you. But you will at least be able to see light from the outside Universe, even though it will be bent and focused by the gravity of the black hole. The Cauchy horizon marks the boundary inside of which you will no longer see light from the outside Universe. Now this might sound reasonable enough at first sight, but don't be fooled. Black holes are such eerie places that nothing is straightforward. One of the bizarre predictions of the mathematics of black holes is what happens to the light you see from the outside Universe as you fall closer towards the Cauchy horizon. Because your time is running more and more slowly, time outside is speeding up until, at the Cauchy horizon, time outside is running infinitely fast and you would literally see the whole future of the Universe flash before you at the instant you pass the horizon. I find this perversely apt; just when you would expect to see your whole past flash before your eyes, you see the entire future instead.

Just to make sure I don't offend any black hole aficionados, I should add that, in reality, you would not really have a privileged

---

[1] Singularities are thus more general than zero-sized points. A singularity is anywhere that marks an edge of spacetime. So, in the 2D rubber sheet model, any cut in the sheet constitutes a singularity.

view of the future of the Universe since all the light that will ever enter the black hole has to arrive all at once in a blink of an eye. The light streaming in will be squashed towards the blue end of the spectrum. This is the opposite of what is seen by an observer outside a black hole watching light falling in. In that case, light is stretched (redshifted). As you approach the Cauchy horizon you see light being more and more blueshifted to higher and higher frequency. This also implies that the light is gaining in energy and you will be frizzled by that final burst of infinitely energetic radiation. Sorry. Of course, all this is assuming you have survived the gravitational tidal forces that will be stretching you and trying to rip you apart before you get to the Cauchy horizon.

Let us for a moment put aside these trivial concerns of being turned into spaghetti and then cooked in radiation and look a little more closely at the singularity itself. The mathematics of general relativity seems to suggest that a Kerr singularity is a window to another universe. Instead of Alice falling down an Einstein–Rosen bridge to Wonderland, here is where she can step *through the looking glass*. You see, provided you can make it as far as the singularity itself, you might be able to travel through the centre of the ring (making sure you do not get too close to the sides of course since that is where the 'stuff' of the singularity is). Once you do this you will have left our own spacetime behind for good.

So where would you travel to if you were to leap through this cosmic ring of fire? The answer is that it depends crucially and uncontrollably on the exact path you take through the singularity. One possibility is that you would end up in a different part of our own universe and, since time and space are mixed up, you would almost certainly end up in a different time too. You might emerge in the distant past or the distant future. [Great, you think, here at last is a real time machine.] But aside from all the dangers of jumping into the rotating black hole in the first place, going through a Kerr singularity is a one-way trip. I do not mean that you couldn't go back through the ring from the other side once you had jumped through, but simply that you would not find yourself back where and when you started. Oh, and don't forget there is still the one-way event horizon that stops you from getting out.

So let me summarize the pros and cons of a Kerr black hole as a 'star gate'. On the positive side, you can avoid being crushed to zero size by carefully navigating through the centre of the singularity. The problem with this is that, from outside the event horizon, you cannot see what angle you should enter. Go in from the side (along the plane of the ring) and you will not be able to avoid spiralling in and hitting the ring. A more important difference between the singularities inside rotating (Kerr) and non-rotating (Schwarzschild) black holes is that space and time are warped in different ways. In the jargon of relativity a point singularity is called spacelike while a ring singularity is timelike. A spacelike singularity marks the edge of time (either its beginning, like the Big Bang singularity, or its end as in a black hole) whereas a timelike singularity marks the edge of space, which is how it can serve as a window beyond our Universe.

All in all, it's a shame about those two troublesome horizons really. The event horizon allows one-way travel only, and it shields the singularity from view so that we are not able to choose the correct angle to enter. The Cauchy horizon, on the other hand, is where you get zapped by infinitely blueshifted radiation. What we would really like therefore is to get rid of these horizons, leaving what is known as a naked singularity exposed to the outside Universe. There are a number of ways of (maybe) getting a naked singularity. One is through Hawking radiation, whereby a black hole gradually evaporates until its horizon shrinks away completely, leaving behind the exposed singularity. But this is still highly controversial and many physicists believe that when a black hole evaporates completely nothing is left behind. In any case, this is only likely to happen to very tiny black holes and it is no good waiting around for a rotating supermassive one to evaporate. Such a black hole might, however, be stripped of its horizons in a different way. You see, the faster a black hole is spinning the further out its Cauchy horizon will extend and the closer it gets to the outer event horizon. Spin it fast enough and the two horizons overlap, and, at that instant, the mathematics predicts that they cancel each other out and both will disappear.

A naked singularity might also form from the collapse of a highly non-spherical mass, but this option is also speculative since such shapes are not likely to exist in the real Universe. The prediction that this type of naked singularity might form comes from complex computer simulations that astrophysicists have studied.

I should remind you of course that most of what I have said so far in this chapter is based on theoretical predictions and speculations anyway. Physicists do not believe that we will ever be able to go through a naked Kerr singularity and travel to another universe or even to the other side of our own universe. Part of the reason for their scepticism (and nervousness) is that if we could we would also be able to use it as a time machine and, as we saw in the last chapter, that is not an option many physicists are even prepared to consider.

But there are a number of practical difficulties that look likely to make the whole idea of travelling through such ring singularities as impossible as trying to go through an Einstein–Rosen bridge. To begin with, it doesn't seem likely that any black hole could spin fast enough to throw off its horizons. And very recent research seems to indicate that the Cauchy horizon is so unstable that as soon as you pass through it (even if you are on course to go through the centre of the singularity) you will disturb it enough to turn it into what is known as a null weak singularity, but a singularity nevertheless, and you would be trapped inside.

When science fact met science fiction

Our current understanding of black holes would seem to indicate that they could never be used in practice as windows or bridges to other universes, or to other parts of our own universe, even if we could ever get to one. They remain a fertile subject for science fiction writers who are not usually deterred by the objections of physicists or even experimental evidence. However, not all science fiction writers disregard the latest findings and predictions of the physicists. Many sci-fi authors are themselves professional scientists and would require of their stories that they at least did not blatantly flout the laws of physics. This was the case in 1985 when the celebrated astrophysicist, author and TV personality Carl Sagan was writing his novel *Contact*. In the story, which has recently been made into a movie, humans make contact with an advanced alien civilization using a tunnel through hyperspace (a wormhole[2]) that links two distant parts of the Galaxy through which the heroes of the story travel. Sagan was aware of the possibility of an Einstein–Rosen bridge or a Kerr singularity for this purpose but wanted his story to be as realistic as possible and needed to get his facts straight. After all, despite the whole idea of fiction being that we can make things up as we go along, being a trained scientist he was determined to only include what was considered at least *possible* by general relativity.

Sagan therefore sent an early draft of the manuscript to his friend Kip Thorne in the Theoretical Astrophysics Group at California Institute of Technology. Thorne is one of the world's leading experts on general relativity and Sagan hoped that he could at least come up with a suggestion or two based on the latest scientific ideas that would add credence to the story. Neither man was prepared for what was to follow. Sagan's request aroused Thorne's curiosity and, with the help of his PhD student Michael Morris, Thorne decided to tackle the problem from an original angle. To understand his approach I should explain what Einstein's equations of general relativity roughly look like. On one side of the equations is information about mass and energy while the other side of the equations describes the curvature of spacetime in the presence of this mass and energy—suffice it to say that the equations are far richer and more complex than his special relativity equation $E = mc^2$. Usually, physicists will start by defining the mass and energy content of a particular region of spacetime, such as a star, then solve Einstein's equations to find out how the surrounding spacetime is affected and what properties it might have. Thorne began thinking about whether wormholes were allowed in theory, but he didn't follow the

---

[2] Note that I have so far avoided using the term 'wormhole' to describe a link between two black holes, whether rotating or not. Instead, I have stuck with the names Einstein–Rosen bridge and Kerr singularity. I do this for a reason that will become clear shortly.

traditional approach. After all, he was well aware of the problems that plagued the usual solutions for black holes, such as event horizons, tidal forces, unstable singularities, tunnels that pinch shut before you can get across and so on. Instead, he decided to start with a wish list. He knew that for the purposes of Sagan's story the wormhole would have to be stable, constantly open, not have event horizons at either end to allow for two-way travel, not have any singularities and not have any uncomfortable tidal forces that would kill any traveller before they could enter. He then set about, with his colleagues at Caltech, to (mathematically) design the shape that spacetime must have to satisfy all his requirements. To his surprise he found that this was indeed possible.

Thorne realized he could design just the sort of wormhole Sagan was looking for. It turned out to be possible in theory to have a link between two parts of the Universe that looked, schematically, just like Wheeler's quantum wormholes of thirty years earlier. But this time the tunnels would be large enough for humans to travel through in a spacecraft without feeling any discomfort. For instance, a traveller could enter one mouth of the wormhole near Earth and within a short time he or she would emerge from the other end on the opposite side of the Galaxy. The traveller would then be able to return through the wormhole and report back. This 'connection' was thus dubbed a 'traversable wormhole' to distinguish it from non-traversable ones like the Einstein–Rosen bridge. From now on, when I refer to such structures I will simply call them wormholes, implying the traversable variety.

Such a wormhole is shown in figure 8.1 in which space is depicted as a two-dimensional sheet. The two entrances into the wormhole are known as its mouths, while the neck (or handle) in between them is referred to as the wormhole's throat. A difficult concept to grasp is that, while the distance through normal space between the two mouths of the wormhole may be arbitrarily long (say a thousand lightyears), the length of the wormhole tunnel itself may be arbitrarily short (a few kilometres or even metres). This is not apparent from figure 8.1 where it looks like the path through the wormhole is actually longer than the one going straight across. However, you must remember

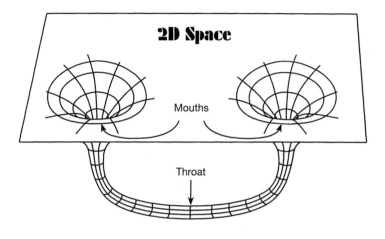

2D Space

Mouths

Throat

Figure 8.1. A wormhole joining two regions of a 2D space.

that the wormhole is really a connection between two regions in curved four-dimensional spacetime which is impossible for us to visualize.

It is also important to appreciate that Thorne's wormhole is not formed from black holes, nor does it have event horizons. So presumably we cannot expect to find one lying about in the Universe. If so, how would we go about constructing one ourselves? First of all, and before you get too excited, building a traversable wormhole is not a job for twentieth or even twenty first century technology. It may indeed never be possible. But since this chapter is dealing in speculation, allow me to speculate. One way of creating a wormhole would be to enlarge a quantum wormhole. Down at the very tiniest length scale and trillions of times smaller than atoms, is what is known as the Planck scale where the concept of length loses its meaning and quantum uncertainty rules. At this level all known laws of physics break down and even space and time become nebulous concepts. Any and all conceivable distortions of spacetime will be popping in and out of existence in a random and chaotic dance which is going on all the time everywhere in the Universe. Terms such as 'quantum fluctuations' and the 'quantum foam' which are used to

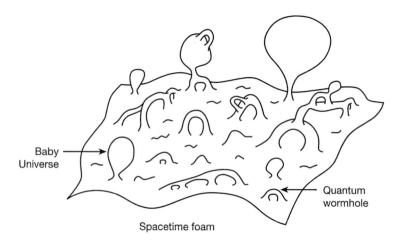

Baby Universe

Quantum wormhole

Spacetime foam

Figure 8.2. Quantum fluctuations. At the very tiniest scale imaginable, spacetime is no longer smooth. Many different shapes will be boiling up and disappearing again.

describe this chaotic activity certainly do not do it justice. This is where Wheeler's tiny wormholes will exist fleetingly before disappearing, and spacetime is said to be 'multiply connected' as in figure 8.2. The trick would be to somehow capture one of these quantum wormholes and pump it up to many many times its original size before it has a chance to disappear again.

We do not yet understand how this could be done in practice. But there might be a hint in the right direction from the way the Universe evolved in the first fraction of a second after the Big Bang. Recall from Chapter 3 that most cosmologists now believe our Universe underwent a short period of very rapid expansion known as 'inflation'. It is thought that this was the mechanism that caused the tiny quantum fluctuations to expand dramatically to become large scale irregularities, or 'ripples', in space. These in turn provided the variations in matter density necessary to produce the galaxies. If the inflationary model is correct then the space containing our whole Galaxy, with its billions of stars including the Sun, was once just a quantum fluctuation much tinier than an atom. If we understood the mechanism that caused inflation we might be able to harness it to inflate our very own wormhole up from the Planck scale to the astronomical scale.

It is clear that, however inflation worked just after the Big Bang, it must have opposed the inward pull of gravity by providing an outward pressure (or antigravity) that would cause space to stretch and the Universe to expand. This idea should sound familiar. It is the work of Einstein's cosmological constant which he first proposed to stabilize the Universe against collapse, and which has been in and out of favour among cosmologists ever since. If we were able to apply such 'negative pressure' to a tiny region of space and cause our own controlled mini-inflation of space we might produce, among other 'things', a wormhole. This means of course that such wormholes may have been created naturally in the Universe. Even so, it is highly unlikely, although not impossible, that some might still be around today as they would have very quickly collapsed.

Of course if naturally occurring wormholes do exist then apart from the difficulty of actually finding one (or one of its mouths at least), we would have no control over where it might lead to. We would just have to try it out and see. The alternative to finding a ready-made wormhole, either a tiny one that we would have to inflate or one left over from the Big Bang, would be to start from scratch and manipulate spacetime ourselves. Even by the speculative standards of this discussion it would appear to be highly unlikely that this would ever be possible. Of course, researchers in 'wormhole physics' are not currently concerned with how to make wormholes since the field is still in its infancy and they are more interested in what their properties are. Scientific papers on this subject often start with phrases such as: "We consider a traversable wormhole joining two asymptotically flat regions of spacetime...", which basically means "Take one wormhole...". They then go on to work through the complex equations of general relativity. Because of the highly theoretical and speculative nature of wormhole physics, such papers often talk about 'cutting and pasting' two regions of spacetime together as a way of creating a wormhole. This conjures up an image of using scissors and tape on a spacetime treated as a 2D sheet of paper. Even the theoretical physicists who write these papers often have this sort of simple image in mind.

As for Kip Thorne, he is no longer as interested these days in wormholes as he was in the late 1980s when his papers started the whole field off. Since then, and throughout the 1990s, many serious and highly technical papers have appeared in the leading scientific journals dealing with wormholes of all shapes and sizes, and interest in the subject shows no signs of abating just yet. Today, it is fair to say that the best known expert on wormholes is Matt Visser at Washington University in St Louis who has written the first textbook devoted to the subject.

Visser has compiled a whole taxonomy of wormholes. He has shown that wormholes come in different phyla and species. The phylum of interest here is known as Lorentzian wormholes (based on the way spacetime is warped to give rise to the wormhole). Lorentzian wormholes are then divided into two species: permanent and transient—we are naturally interested in permanent ones. Each of these species consists of two subspecies depending on whether it is a wormhole that connects two different universes (known as an inter-universe wormhole) or two, possibly distant, regions of the same universe (an intra-universe wormhole). Each of these subspecies is then divided into macroscopic and microscopic varieties. The 'macroscopic' tends to mean traversable, while 'microscopic' implies quantum wormholes of the type first studied by Wheeler. In general of course, Wheeler's quantum wormholes are of the transient type since they pop in and out of existence according to the rules of quantum mechanics. But, due to quantum mechanical uncertainty, it may be that wormholes of the permanent variety (by which I mean having the right spacetime curvature that allows them to last much longer than normal) might occasionally be created.

Wormholes—keeping the star gate open

Of all the properties of wormholes, the one that has been aired the most is the issue of their stability. You might find this surprising given that we are not even sure how to make one in the first place, and in a sense you would be right. But wormhole physics is all about what is possible—or rather what is not impossible. It is enough that wormholes of the type first proposed by Thorne *can* exist theoretically. How they are created is of secondary importance. What is not yet clear however is whether the wormhole can be kept open long enough for someone moving at a comfortable speed (considerably less than the speed of light) to get through.

One of the conditions that Thorne wished to impose on his traversable wormhole was that it would not pinch off quickly like an Einstein–Rosen bridge or snap shut as soon as we tried to go through like a Kerr singularity. However, he discovered that this would not be an easy matter. The throat of the wormhole would not stay open of its own accord and needed a lot of help. You might think that this would be the least of his problems since we can imagine having to erect some sort of scaffolding within the wormhole that would be of such strength that it could withstand the immense gravitational forces trying to close it. This would obviously be way beyond our technological capabilities, but *not impossible*. Unfortunately, it turned out that no known matter in the Universe could fulfil Thorne's requirements. He realized that the only way his wormhole would stay open would be if it was threaded with a very strange kind of material that would have to have negative mass! What could this mean? How can something have a mass that is less than zero? Technically, it is said to have negative energy since mass and energy are interchangeable, which is just as preposterous. In typical scientific understatement, Thorne dubbed this material 'exotic'.

A common confusion that many people have when they hear about this is that it is the same thing as antimatter. Far from it. Antimatter is a piece of cake compared with exotic matter. Antimatter has sensible positive mass and is every bit the same as normal matter in its effect on spacetime. The difference between matter and antimatter is that they have other opposite properties, such as electric charge. So, just as a subatomic particle such as an electron is negatively charged, there exists its antimatter equivalent, the positron, which is identical to the electron in every way apart from being positively charged. If a lump of matter is brought together with a lump of antimatter they will mutually annihilate in a burst of pure energy. But an isolated lump of antimatter will fall towards the Earth obeying the laws of gravity just like normal matter. Exotic matter, on the other hand will, if dropped, experience a force of antigravity repelling it away from the Earth's surface!

So where would you go to buy a sufficient amount of this exotic material to use in your wormhole? Well, we do know how to make a very tiny amount of negative energy. Not much, but it's a start. Work on traversable wormholes has rekindled interest in an obscure yet fascinating, and experimentally proven, effect discovered by the Dutch physicist Hendrik Casimir in 1948. It involves a property of what we would consider to be completely empty space.

If all the air is pumped out of a chamber then we say that we have a vacuum, meaning that there is no matter inside and hence, I hope you'd agree, zero energy. But down at the quantum level, even the empty vacuum is a busy place. I recommend at this point that you go back and reread the section in Chapter 4 entitled 'Not so black after all' where I discuss Hawking radiation. That is where I describe how particles and their antimatter partners are continuously popping into existence from nothing before quickly disappearing without a trace. Casimir showed how to harness this process to extract energy from the vacuum even though it has nothing to give.

As most of us know only too well, if we borrow money from a bank we must soon pay it back. The rules of quantum mechanics, as expressed within the Heisenberg uncertainty principle, operate in a similar way. But unlike a bank loan where we are free to choose the period over which we make the repayments, the uncertainty principle is rather more strict. It states that energy can be borrowed from the vacuum provided it is paid back very quickly. The more energy that is borrowed, the quicker the dept must be repaid. Now consider what is going on in a vacuum if we could zoom down to the microscopic level. Among the myriad of subatomic particles that are forming from this borrowed energy are photons (the particles of light). What's more, photons of all energies are being created, with the higher energy ones, corresponding to short wavelength light, being able to stick around for much less time than the lower energy, longer wavelength, ones. Thus at any given moment, the vacuum contains many of these photons (and other particles) and yet will have an average energy equal to zero since each particle has only temporarily borrowed the energy needed for it be created.

Casimir showed how the vacuum can be coaxed into giving up a tiny amount of its energy permanently. This is achieved by taking two flat metal plates and placing them up close to each other inside a vacuum. When the distance between the plates is not equal to a whole number of wavelengths, corresponding to photons of a particular energy, then those photons will not be able to form in the gap because they will not fit. This is a rather difficult concept to appreciate, since we must consider both the wave nature of light (wavelengths) and its particle nature (photons) at the same time. Nevertheless, the number of photons forming in the vacuum between the plates is less than the number on the other side of the plates and it will therefore have a lower energy. But since the vacuum outside the gap has zero energy already then the region between the plates must have less than zero (or negative) energy. This causes the two plates to be pushed together with a very weak force that has nevertheless been experimentally measured[3]. Unfortunately, the amount of negative energy that can be made in this way is very tiny and is nowhere near enough to keep a wormhole open. But it's a start.

In keeping with the spirit of this chapter, I am not proposing that the Casimir process will one day lead to enough exotic matter to line a wormhole's throat, but rather that such negative energy material, albeit very tiny and extracted from empty space, is not ruled out by the laws of physics. In fact, some physicists have proposed that there might be a way of squeezing the vacuum and pumping energy out of it in a more systematic way, but this is by no means clear yet. Just to give you a feel for the amount of exotic

---

[3] The experimental verification of the Casimir effect is still a bit controversial.

material that is needed, Matt Visser has calculated that we would need exotic matter equivalent to the mass of Jupiter just to hold a one metre wide wormhole open.

Another way of getting hold of exotic material is from something called cosmic string. This is material that might have been left over from the Big Bang but whose existence is highly debatable. It should not be confused with the string of superstring theory which I will discuss further in the last chapter, but is much more impressive. Cosmic string would either be in the form of a loop or would stretch right across the Universe (and thus may be infinitely long if the Universe is infinitely large). Either way, this is string that doesn't have an end! Its diameter is much less than the width of an atom yet it is so dense that just one millimetre of it would weigh a million billion tonnes. The hope would be that if the Universe went through an inflationary period, driven by antigravity due to a non-zero cosmological constant, then the state of the Universe at that period may have been frozen within the cosmic string. The string would therefore contain exotic matter, or whatever it was that caused the antigravity driven inflation during that time. If we could find such string in the Universe it would be just right to thread through our wormhole.

Visiting a parallel universe

So far, I have only discussed intra-universe wormholes which would connect two distant points in our own universe. But Wheeler's quantum wormholes might also connect us with a parallel universe. These are referred to by Hawking as baby universes since they would be like bubbles which form and grow out of the quantum foam in our Universe and connect to ours via a wormhole like an umbilical cord. Such a baby universe might itself then start to expand within higher dimensional hyperspace and, if the wormhole connection is broken, be forever separated from ours. In fact, if this is true then our own universe might itself have popped out of the quantum foam of yet an earlier universe.

Thus, in the same way that we would inflate an intra-universe wormhole up from the quantum level, we can also think of inflating such an inter-universe wormhole that connects us with a neighbouring universe. If it turns out that we live in a closed universe, then one can imagine a distant future in which the Universe is collapsing towards a Big Crunch. If humans are still around at that time they will want to be able to escape the crush of the final singularity by jumping through a wormhole into a younger universe which had the same properties as our own.

9

HOW TO BUILD A
TIME MACHINE

Recipe for dragon stew: First, find a dragon...
Matt Visser, *Lorentzian Wormholes*

Having come this far, you are finally in a position to appreciate the physics that needs to be in place if we are to construct a time machine. I have discussed Einstein's two theories of relativity, both the special theory in which time and space are united into four-dimensional spacetime, and his general theory in which spacetime is warped and twisted in the presence of matter and energy. Both theories are going to be needed in this chapter. I have discussed the nature of time and looked at the sort of problems that we must overcome if we insist on the possibility of time travel into the past. Now it's pay-off time.

I will put on hold all the (quite valid) objections to time travel for the time being and adopt the pragmatic, and highly optimistic, view that as long as time travel is not forbidden by the laws of physics as we understand them today then there is hope. I will show how we might go about building the simplest possible time machine. Don't take this to mean that I have climbed down off the fence in favour of time travel, but rather that I am leaning dangerously over to one side (for now).

Time loops

I read an article in a Sunday paper recently with the headline 'Could Einstein have been wrong after all?'. "Oh no," I thought, "another crazy idea trying to disprove special relativity." For me, to read that special relativity has been proven wrong would be equivalent to reading that it has been discovered that the Earth is flat after all. Both would be quite preposterous given everything we know. But it was general relativity that was being discussed in the article and, far from being under threat, was alive and well. It was just the headline that was misleading.

Many physicists regard general relativity to be the most beautiful scientific theory ever discovered. Its beauty lies in the simplicity, elegance and richness of its mathematical equations. I admit that only a tiny fraction of the human population can appreciate this beauty because they have had years of training. For most people it is just a bunch of Greek symbols. But then I have never been able to appreciate or understand cubism as an art form. Anyway, as well as being pleasing to theoretical physicists, general relativity has been confirmed by experimental evidence time and time again. However, in almost all these cases it has only been in what is called the 'weak field limit' (i.e. weak gravity). It has yet to really prove its mettle in situations where it departs radically from Newtonian gravity.

The newspaper article I mentioned described a new type of experiment that, it is anticipated, will confirm yet another prediction of general relativity known as gravity waves. The first paragraph in the newspaper article was in fact stating that if such gravity waves *were not found* then general relativity would be in trouble. However, such is physicists' faith in the theory that those working on these new experiments fully expect to find what they are looking for very soon. Unfortunately, a headline proclaiming: 'More experimental proof that Einstein was right is just round the corner' is simply not as newsworthy.

Gravity waves are disturbances, or ripples, in the fabric of space which are caused by the motion of a massive object. Think of the trampoline model of space. When you stand in the centre your weight makes a dent in the canvas. This is the simple view of how mass affects the space around it. If you then jump up and down you will make the canvas vibrate and, provided it has a very large area, these vibrations would travel outwards in the same way that ripples spread out on the surface of a pond when a stone is thrown in. Likewise, the motion of massive objects such as the collapse of a massive star into a black hole will send out ripples not *through* space but *of* space that will affect any objects in their path. The hope is that experiments on Earth will be able to detect the effect these gravity waves have on sensitive equipment which will be ever so slightly stretched and squeezed as the waves pass through.

Of course gravity waves have nothing to do with time machines. I mention them as an example of an unambiguous prediction of general relativity that has yet to be confirmed experimentally. However, general relativity is so rich that it also allows (theoretically of course) other, more exotic spacetime shapes to exist which we are not nearly so confident about. One of these, of relevance to this chapter, is called a *closed timelike curve*. This is a circular path, or route, through warped spacetime in which time itself is bent round in a circle. If you were to follow such a path it would seem to you that you were travelling through ordinary space. If you were to check your watch at any time during this round trip you would see it running forwards as normal. However, you will, after some time had elapsed for you, eventually arrive back at the same place *and* time that you started from according to a clock that had been left there. Such a path would require you to be travelling into the past for part of your journey. Of course if you are travelling back in time you might as well get back to where you started from *before you set off*, otherwise you will not have gained anything from the trip. Any further warping of spacetime will cause the time loop to take you back into the past.

Thus 'closed timelike curve' is Jargonese for 'time machine'. I shall refer to them here simply as 'time loops'. It has long been known that general relativity allows for the existence of time loops, but whenever one popped up in the mathematics it was usually disregarded on the grounds that the initial assumptions that were fed into the equations were unreasonable. What party-poopers physicists are. Unfortunately, this attitude is justified and there are plenty of other examples to illustrate why. Take the following simple one. If you are told that a square has an area of 9 square metres then you deduce that it has sides of length 3 metres, since the area of the square comes form 3×3. However -3×-3 also gives 9 (remember that a negative times a negative gives a positive). But, we would never talk about the side of a square having a length of -3 metres and so we ignore this option because it is unphysical. Mathematical equations that describe the real world often give, along with the correct answer, such unphysical, or nonsensical answers, which should be ignored. For the vast majority of physicists working on general relativity, time loops fall into this category. They are considered unphysical because of all the problems associated with time travel into the past.

In recent years, however, some physicists have been more reluctant to dismiss time loops so quickly and they have become a fashionable field of study. As we shall see, this is in part due to Kip Thorne's work on wormholes. However, despite the ease with which time loops can be produced as solutions of Einstein's equations, physicists are still undecided whether they can really exist in our Universe.

The first solution of Einstein's field equations of general relativity that described a spacetime containing time loops was due to W J van Stockum in 1937. However, a connection between this strange mathematical solution and the possibility of using it to describe time travel was not appreciated until much later. The van Stockum solution required an infinitely long cylinder of very densely packed material spinning rapidly in empty space; not the sort of thing you are likely to come across by accident unless you are on board the Starship Enterprise. General relativity predicted that the region of spacetime surrounding the cylinder would be twisted around it and could contain a time loop. But the infinitely long cylinder was understandably dismissed as too unreasonable to be taken seriously. What is more, the mathematics predicted that even spacetime very far from the cylinder would have strange properties, proving that such a cylinder of matter could not exist in our Universe or we would be able to see its effects locally even if it were on the other side of the Universe.

Time loops really hit the scientific headlines with the work of Kurt Gödel in 1949. In a classic paper, he described mathematically an abstract universe that would contain time loops. However, Gödel's universe differed from the one we inhabit in the way it maintained its stability against the inward gravitational pull of its matter. Instead of expanding, as ours does, his universe was rotating. If a space traveller in such a universe were to follow a large enough circular path then she would get back to her starting point before she set off. Time travel!

Although Einstein, who worked in the same building at Princeton's Advanced Study Institute as Gödel, was initially disturbed by this result, he (and most other physicists) soon dismissed the result as being of little relevance to the real Universe which we know is *not* rotating. Even Gödel himself ignored the possibility of time travel because it was so unachievable in practice, not just because his model universe was unlike the real one, but because of the unrealistic speeds required and distance that would have to be covered by a rocket in order to complete a time loop in such a universe. The fact remains, however, that Gödel had come up with a scenario (albeit an unrealistic one) in which no laws of physics were violated and which was entirely consistent with general relativity, but which contained time loops with all the time travel paradoxes they implied. Most physicists believed, and still do believe, that the loopholes in the physics that allow this sort of solution to exist will eventually be plugged up through a better understanding. Until then, Gödel's universe has been relegated to the status of a mathematical curiosity.

The Tipler time machine

By the 1960s and '70s many more theoretical models of spacetimes that contained time loops were discovered by a number of physicists who were studying the properties of the equations of general relativity. All these models had one thing in common. They involved rotating massive objects that twist spacetime around them. The best known work along these lines was due to a young American by the name of Frank Tipler who published a paper in 1974 which caused quite a stir at the time. Tipler had re-analysed the work of van Stockum involving a rotating cylinder, and took it a step further. First he proved mathematically that, to be sure of a closed time loop around the cylinder, the cylinder did indeed have to be infinitely long, made of very dense matter and spin at a rate of thousands of times per second. The biggest problem was, of course, the 'infinitely long' bit, which is easier said than done. So Tipler then went on to calculate what would be needed to build a time machine in practice. He suggested that we might be able to get away with having a cylinder just 100 kilometres long and 10 kilometres wide. The problem was that he could no longer rely on the mathematics to prove that this would be sufficient to warp spacetime enough. And even if a closed time loop could be achieved, the cylinder would have to be fantastically strong and rigid so as to avoid being squashed down along its length due to the enormous gravitational strain it would be feeling. At the same time, it would have to be strong enough to hold together and withstand the enormous centrifugal force trying to fling its matter outwards as it spun at a surface speed of over half that of light. However, he pointed out that these were all practical problems and, anyway, who knows what might be technologically possible in the distant future.

To use a Tipler cylinder time machine, you would leave the Earth in a spaceship and travel to where the cylinder is spinning in space. When you are close enough to the surface of the cylinder (where spacetime is most warped), you would orbit around it a few times then return to Earth, arriving back in the past. How far back depends on the number of orbits you made. Even though you feel your own time moving forward as normal while you are orbiting the cylinder, outside the warped region you would be moving steadily into the past. This would be like climbing *up* a spiral staircase only to find that with each full circle climbed you are on a floor *below* the previous one!

A number of other researchers have also suggested that we may not need infinitely long cylinders to get time loops, and that time travel may even be achieved by orbiting round a spinning neutron star or black hole, provided they were spinning fast enough. Astronomers have already found neutron stars (pulsars) that spin close to the required rate. These are known as millisecond pulsars because their rate of spin is once every few milliseconds (a millisecond being one thousandth of a second). Some claim that we need to simulate a long cylinder, in which case we would need to pile a number of such millisecond pulsars on top of each other, then find a way of preventing them from squashing down into each other and forming a black hole[1]. Other calculations imply that just one rapidly spinning black hole which has shed its horizons, leaving behind a naked ring singularity, is sufficient to provide a closed time loop around it. However, the mathematics for all these wild and wonderful suggestions is far from conclusive.

Cosmic string time machines

One possible way that Tipler's time machine could be realized is by using cosmic string. We saw in the previous chapter how useful cosmic string would be in keeping a wormhole's throat open. Yet again, this might be just the right sort of material we are looking for. It would be infinitely long and would certainly be dense enough. All we would need to do is get it spinning fast enough. This does of course assume that (a) cosmic strings exist, (b) we are able to locate and travel to one, (c) we could find some way of spinning it fast enough and (d) a closed time loop really would form around it.

Even when a cosmic string is not spinning, spacetime around it is distorted in a rather strange way (yes, stranger even than spacetime around a black hole!). Despite the high density of the string, you would not feel an attractive gravitational force however close you were to the string, and spacetime is said to be flat. However, space by itself will be cone-shaped around the string. To see this, consider 2D space for simplicity and a circular patch of this space around the string. It would be as though a wedge

---

[1] Theoretical speculation is getting a little out of hand when we start talking about stacking neutron stars together, but it's fun.

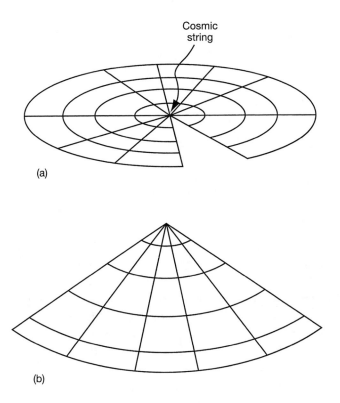

(a)

(b)

Figure 9.1. To model what 3D space looks like near a cosmic string (a 1D line), throw away one dimension and consider 2D space around a (0D) point. If a wedge of space is removed as in (a) and the two edges pasted together as in (b) then space will be cone-shaped.

of space had been removed as in figure 9.1(a) and the two edges pasted together as in figure 9.1(b). This will cause space to be closed round the cosmic string in the shape of a cone. Normally, the circumference of a circle is given by twice its radius multiplied by *pi*. And so, if a piece of the circle is missing, its circumference would be less for the same radius. If you were to travel in a circle of a given radius around the string you would find that the distance you travel to get all the way round and back to your starting point

223

would be shorter than the distance you would cover round a circle of the same radius in normal space (away from the string). Note that the string is depicted by a 0D point in 2D space. It is really a 1D line in 3D space (which I am unable to show since I cannot draw a 4D cone!).

A variation on this cosmic string theme was suggested by Richard Gott in 1991. He provided a way round the requirement that the cosmic string needs to be spinning. Instead he showed how two strings moving past each other at high speed would have the same effect, and a time loop would form around the pair. The problem here is that the two strings would have to be parallel to each other as they passed. So even if cosmic string does exist, we would still have to hope for two strings to just happen to encounter each other at just the right angle. Gott points out that we need not wait for two infinitely long strings to pass each other. The same effect might be achieved if one closed cosmic string, forming a loop, which was an oval rather than circular (like the shape formed by a stretched rubber band) were to collapse in such a way that the two long sections just miss as they fly past each other. Gott himself has pointed out that any closed time loops that might form around two pieces of cosmic string that were not infinite in length would form a black hole and be shielded from the outside by an event horizon, which would, of course, mean that they could never be used.

Unfortunately, Gott's way of achieving a time loop is even more hare-brained than the other schemes on the market since, along with all the ifs and buts I have already mentioned, it requires part of the total mass of the strings to be what is known as 'imaginary'.[2]

---

[2] This is even worse than having negative mass, which is ridiculous enough. The word imaginary implies having a certain mathematical property that involves the square root of a negative number. If you have not encountered this sort of thing before here is a very brief explanation. You know that the square of a positive number is a positive number, and the square of a negative number is also a positive number (since $- \times - = +$). But a number that when multiplied by itself still gives a negative number is known as an imaginary number because it is not like normal (real) numbers and has its own set of rules. Such numbers are useful in many fields of physics and engineering.

A recipe for a wormhole time machine

All the ways I have discussed so far of getting closed time loops have involved warping spacetime around a spinning mass. There is an alternative way that does not involve travelling round a dense massive object, but requires a wormhole instead. Soon after Kip Thorne had shown his friend Carl Sagan how a traversable wormhole might be constructed so as to connect two distant regions of space via a short tunnel, it was pointed out by colleagues that there would be no reason why the wormhole should not also join two different times. After all, it is 4D spacetime that the wormhole is being created in, not 3D space by itself. A simple-minded way of visualizing this is by using the block universe model. Figure 9.2 shows a wormhole through this 3D spacetime which connects two different times. An important point to note here is that the throat of the wormhole is not actually imbedded within the block, but exists in some higher dimensional space outside the three dimensions in which it is drawn. Unfortunately, I ran out of dimensions. But at least it gives you a rough idea what is involved. If your 'now' happened to be at the LATER TIME then travelling through the wormhole would take you into the past. But, equally, someone whose 'now' was in the EARLIER TIME would use the wormhole to travel into the future.

The way I have drawn the wormhole in the block universe suggests that its two mouths open up in specific slices through the block. This might imply that someone whose 'now' is on a slice half way between the two will not see any wormhole mouths. In fact, all slices that come after the EARLIER TIME slice will also contain its wormhole mouth, since they are just that slice at later times. All slices after the LATER TIME slice will thus contain both wormhole mouths. One would be a link into the past and the other to the future.

What makes the block universe model quite hard to appreciate, but which is necessary for time travel to be possible, is that neither time has the right to call itself the real 'now'. Both are equally valid since time does not move in the block universe. This does not stop our subjective feeling that time is moving along. So, to us living within the block universe both slices will move

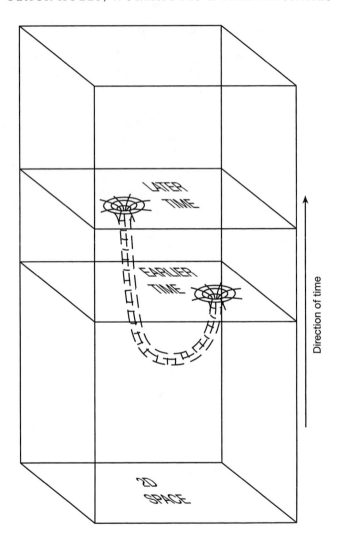

Figure 9.2. A wormhole in the block universe joining two different times.

upwards at the same rate, along the time axis, but there will be people whose present moments correspond to one or other of the two times.

Kip Thorne and his colleagues were even able to show how a wormhole that was not a time machine—in the sense that if you were to go through it you would emerge at the other end at a later time with the same amount of time having elapsed for you inside the wormhole as had gone by on the outside—could be turned into a time machine. What I mean by a wormhole that is not a time machine can be understood from figure 9.2 if the EARLIER TIME and LATER TIME slices were one and the same. Now if you travel through the wormhole you will emerge in the same time slice that you would be in had you not travelled through. The trick in creating the time machine was to make use of an effect in special relativity which you have already met. It involves the idea behind the twins paradox.

Let me first lay out the plan for what is regarded as the easiest way of constructing a time machine (assuming it is possible of course):

How to build a time machine

(1) **Make a wormhole** (*by inflating one out of the quantum foam, or creating one from scratch by warping spacetime*).

(2) **Stabilize the wormhole** (*by keeping it open with exotic matter or cosmic string*).

(3) **Electrically charge one of the wormhole mouths** (*so that it can be moved about with an electric field*) **and load it onto a rocket.**

(4) **Induce a time difference between the mouths** (*by flying off at close to the speed of light with one of the mouths*).

(5) **Turn the wormhole into a time machine** (*by bringing the mouths closer together again*).

Steps (1) and (2) were discussed in the last chapter. But to recap, we really have no idea how this could be achieved. This is why many authors who discuss wormhole time machines usually start glibly with the statement: "Take one traversable wormhole". Since I have no more to add to this I will do likewise and assume that we already have a stable wormhole which, for convenience, has mouths large enough for humans to walk through. The two mouths could initially be side by side in our wormhole laboratory. Step (3) enables one of the mouths to be transported from the wormhole laboratory and into a waiting rocket. For steps (4) and (5) we can forget the wormhole mouth is even in the rocket. All we need to do is have the rocket fly around at close to the speed of light for a while and the special relativistic effect of time dilation will do the rest.

Remember the twins paradox story? When Alice heads off from Earth in her high speed rocket and travels around for a while she will, on returning to Earth, find that more time has elapsed there than she can account for. She will return younger than her twin brother, Bob, because she has effectively fast forwarded into the future. This time, Alice will take with her one mouth of a wormhole and cause a time shift between the two mouths.

The following description of what this would be like is similar to the one discussed by Kip Thorne in his book *Black Holes and Time Warps*, but with some modifications. Figure 9.3(a) shows Bob in the wormhole laboratory looking through his end of the wormhole at Alice. Through the wormhole she is only a few metres away, but she is in fact sitting in her rocket which can be seen outside the window on the launch pad. Figure 9.3(b) shows the view from Alice's wormhole mouth which is secured inside the rocket.

Alice and Bob agree that she will fly off on a trip round the Solar System travelling at her rocket's cruising speed of a hundred thousand kilometres per second (or one third of the speed of light) and return to Earth after exactly two weeks. Let us say she departs on a Wednesday. As she speeds away from Earth, the distance between her and Bob through the wormhole remains constant (see figure 9.4), even though he can see her rocket through his telescope receding from Earth at one third of the speed of light. Bob is able to chat to her and even pass freshly brewed cups of coffee to her through the wormhole each morning. More importantly, they will be counting down the days together. At all times during the journey, their watches will be in agreement, since they make sure they remain synchronized through the wormhole.

Two weeks later, with both Bob and Alice agreeing that it is Wednesday, Bob watches his sister through the wormhole as she

(a)

Figure 9.3. (a) Bob can see Alice through the wormhole which provides a short cut between the two of them regardless of how far apart they are in 3D space. She is, in fact, sitting in her rocket which can be seen out of his window.

nears the end of her journey and manoeuvres her rocket through Earth's atmosphere before finally landing it back on its pad. Bob goes outside to watch her land, but the sight that greets him as he steps outside is quite a shock. The launch pad is empty, rocketless. He pulls himself together, dashes to the observatory and aims his

(b)

Figure 9.3. (b) Alice can see Bob through her end of the wormhole.

telescope at the patch of sky where the rocket would have come from. Such is the resolving power of the telescope that he is able to pick out Alice's rocket just flying past Neptune on its journey towards Earth. He calculates that at her current velocity she will not reach Earth till tomorrow!

Being a scientist—he works in a wormhole laboratory after all—it quickly dawns on Bob that this is exactly what he would expect. He can explain what is happening by appealing to special

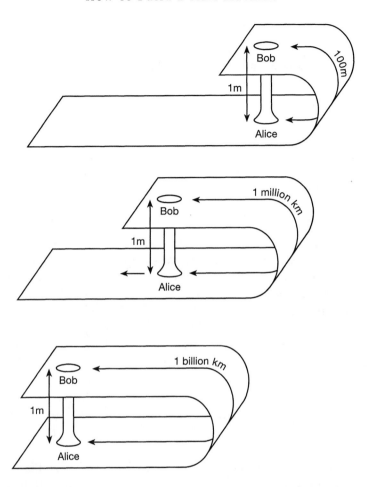

Figure 9.4. The two wormhole mouths can be arbitrarily far apart through normal space and yet remain close together through the wormhole.

relativity. He runs back to the wormhole laboratory to tell his sister. Once inside, he looks through the wormhole to see Alice just completing her final checks of the rocket's controls and getting ready to open the door to climb out. He calls through the wormhole congratulating her on yet another perfect landing, then proceeds to inform her that she hasn't actually landed yet!

She waves back to him. "What do you mean I haven't landed yet? You just saw me land. I hope there's some coffee waiting, there's something about that exotic matter in the wormhole that really ruins coffee."

"Wait a minute Alice," shouts Bob a little frantically now, "I mean it. I think you've moved into a different time frame to me. I know I saw you land the rocket through the wormhole, but out there"—he waves vaguely in the direction of the launch pad through the window—"you are still in the outer solar system. Your rocket is certainly not out on the launch pad. In fact, you are not due back till tomorrow!" Alice, not surprisingly, is far from convinced. She can see that Bob is serious, but then there seems to be nothing illogical as far as she can see. She tries again: "Look, we both agree that it is Wednesday. In fact we've both been counting off the days together. What's more our watches are still synchronized. Therefore we must both be in the same time frame. And, believe me, I *did* just land this rocket."

But Bob is not listening any more and is deep in thought. A few minutes earlier he was convinced he understood why Alice always claimed on returning from her travels to have been away for a shorter time than had passed by on Earth. That was just special relativity at work. But this damned wormhole really seemed to be screwing things up. Then, just for a moment, the fog lifts and he understands. He starts blurting it out before he becomes confused again.

"Alice, let's just say that we didn't have the wormhole. I wouldn't know that you had landed. In fact for me, in Earth time, your trip really would take fifteen days and I wouldn't see you till tomorrow. But for you, rocket time, the journey will only take fourteen days. Less time will have elapsed for you because of your high speed. So you land the rocket on Wednesday according to rocket time, but Thursday Earth time. You have moved one day into the future." He pauses to make sure she follows. "Go on," she says excitedly.

"Well, it doesn't matter that our times are synchronized through the wormhole. Throughout the journey, you have been steadily dragging your wormhole mouth into the future of Earth time. I know it seems like Wednesday to you. It is, inside the rocket. But now that you have landed back on Earth I am afraid you have to abide by Earth time."

"You sound like an air stewardess" she laughs. "Thank you for flying with us, and please adjust your watches to local time, where it is Thursday."

"Yup. The Earth you have landed in is my tomorrow."

"I prefer to think that I am in the present if that's OK with you bro."

"Fine. If you insist on being in the present then what you see when you look through the wormhole is one day in your past. You are seeing a time that happened yesterday for you, when you were still flying back. But I can equally claim to be living in the present and I am looking through the wormhole at what will happen tomorrow. At least I know that you will land the rocket safely."

"What now?" asks Alice.

"Well, the wormhole has been converted into a time machine. Not a very versatile one I'll admit but one which will constantly connect two times one day apart."

Alice and Bob can now use this two-way wormhole time machine as often as they like. He can step through it to join his sister in Thursday, or she can join him in Wednesday. They can buy Thursday's newspaper, look up the previous evening's National Lottery result, climb back into Wednesday and pick the winning numbers.

Of course Alice could join Bob in Wednesday and they can then wait till Thursday, and both go outside and watch Alice land the rocket! The Alice in the rocket will at that moment be chatting to Bob of Wednesday and will eventually climb through the wormhole to join her brother and become the Alice waiting outside. So for a while there will be two Alices. Presumably had Alice looked out of the rocket before she climbed through the wormhole she would have seen herself, and another Bob.

I have deliberately avoided any time travel paradoxes in this story. But if you are looking for trouble, they are very easy to find. To give you an example, what would happen if the Alice and Bob who are standing outside watching the rocket land on Thursday were to go over to it and climb aboard (from outside the rocket and not via the wormhole), and attempt to stop the Alice in the rocket from going through the wormhole? Not only must they fail because she does go through, but they cannot even make contact with Rocket Alice since Outside Alice has no memory of such an encounter with herself when she was Rocket Alice!

This is just the no choice paradox rearing its ugly head, and we must appeal to one or other of the two methods of resolving it that I discussed in Chapter 7:

(a) If Outside Alice cannot remember seeing herself when she was in the rocket then she will clearly be forbidden (somehow) from interacting in any way with Rocket Alice. Physicists refer to such a scenario as an inconsistent solution.

(b) The Universe splits into two the moment the wormhole becomes a time machine.

You may well have taken the above delicious nonsense with a large dose of salt. However, I should point out that not only does my story nowhere violate any laws of physics, but too much salt is bad for you, and ruins the flavour. So how likely is it that such a scenario could become a reality in the distant future?

As I explained at the end of the last chapter, steps (1) and (2) in the 'How to build a time machine' box may never be realized anyway. But if they are and we can make a stable traversable wormhole, are there any other obstacles?

Step (4), which involves inducing a time shift between the two wormhole mouths by moving them apart at very high speed, is only a problem if you do not think it will ever be possible to build a rocket that can travel at near light speed. Of course that is not the only way to induce the time shift. We could, if we had access to a strong enough gravitational field, use the general relativistic time dilation effect to slow the time down at one end of the wormhole. This could be achieved by taking one wormhole mouth on a trip round a black hole a few times. Of course the orbit does not need to follow a closed time loop in this case since we do not require the orbiting wormhole mouth to travel back in time. All we need is for time to be slowed down relative to the other mouth which is far away from the black hole.

Insurmountable problems?

There have been a number of objections to the wormhole time machine plan. All have been based on serious calculations which have shown one or other of the steps to be an obstacle that we could never overcome. The most serious of these has been that, even if a traversable wormhole could be built, and a time shift induced, the last step of bringing the two mouths close together—which you might have thought would be the easiest—would in fact cause the wormhole to be destroyed. It is expected—though no one is sure yet—that as soon as the wormhole becomes a time machine, light which has travelled through it will be able to get back, through normal space, to the mouth it entered *before* it entered. It will then be able to enter along with its original version, thus doubling up its energy. But if twice as much radiation goes through and can get back to the entrance before it enters then it should be four times as intense, and so on. In fact, calculations show that the instant the two mouths (with their different time frames) are brought close enough together for a little of the light leaving the exit mouth to get back and go through the entrance mouth before it went in, an infinite amount of light will instantly have built up by flowing through the wormhole and will either collapse the throat of the wormhole or cause its two mouths to explode in a burst of energy. This light (or electromagnetic radiation) will always be a problem as it is produced by the vacuum itself, and is therefore referred to as vacuum fluctuations.

One wormhole expert, Tom Roman, has shown how a wormhole time machine can be constructed without having to follow step (5) of bringing the mouths together. Instead, once a time shift had been induced between the two mouths and while they are far apart, the two mouths of a second wormhole, which do not need to be time shifted with respect to each other, could be placed adjacent to those of the first, as in figure 9.5. Now you

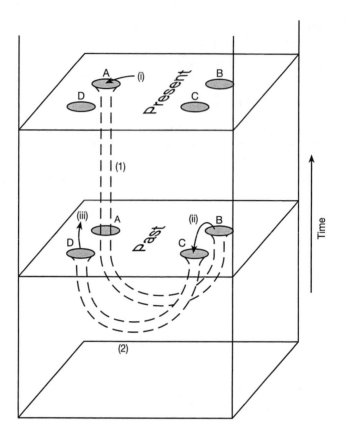

Figure 9.5. A Roman time machine using two wormholes. Having induced a time shift between mouths A and B of wormhole (1) you need not risk destroying it by bringing them close together again. Instead, use wormhole (2) to get you back to your starting point. Step (i): go through A. Step (ii): come out of B and go into C which is next to it. Step (iii): come out of D which is next to your original entry point at A, but in its past.

can go through the first wormhole and then use the second one as a short cut back to your starting point, returning before you originally left. The vacuum fluctuations which pile up and destroy a single wormhole when the mouths are brought together are now

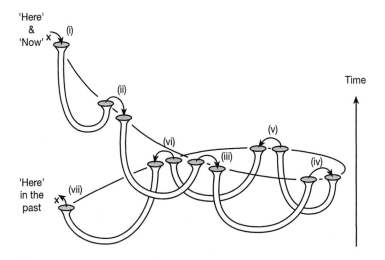

'Here'
&
'Now'

'Here'
in the
past

Time

Figure 9.6. A Roman ring time machine using many wormholes, each of which has a small time shift between its two mouths. The time shifts build up while the wormholes take you on a round trip back to your starting point, in the past.

kept under control. However, all calculations so far carried out show that the equations of general relativity still give nonsensical answers as soon as a time loop is formed from this so-called 'Roman' configuration of wormholes. Matt Visser suggests that we could place a whole series of wormholes in a ring. Each wormhole would have a small time shift but not enough for it to be used as a time machine by itself because, even if we were to go through it, the distance between the two mouths in normal space would be too far for us to get from the exit mouth to the entrance mouth before we entered. But the combination of all the wormholes such as is shown in figure 9.6 might work.

Stephen Hawking is convinced that nature forbids time loops and time travel. He has come up with what is known as the chronology protection conjecture which states simply that time travel into the past should not be allowed in physics. This has yet to be proved mathematically but so far has equally not been disproved. If the chronology protection conjecture does turn out to be a law of nature then time travel would be ruled out for good. The Roman configuration of wormholes (whether with two or many) also appears to have problems, with the mathematics not giving a definitive, or even sensible, answer.

Latest results indicate that an even earlier stumbling block is the exotic matter needed to keep a traversable wormhole open in the first place. A number of researchers claim that their calculations rule out the possibility of ever getting enough exotic matter to hold open any wormhole bigger than a quantum one.

So what are we to believe? Can wormholes ever be built? Can they form time machines? Can time loops be formed at all in the Universe? You will have gathered from everything I have said in the last few chapters that the jury is still out on all these questions. We simply do not know for sure yet. In fact, apart from the obvious problems of how to keep a wormhole open and how to transport it, there is a more basic theoretical stumbling block.

Quantum mechanics describes the behaviour of the world of the very small, whereas most phenomena that are described within the framework of general relativity tend to involve immense expanses, leviathan masses and titanic forces. Stars, black holes, galaxies, even the whole Universe, all rely upon Einstein's description of gravity and how it affects space and time. They are far removed from the microscopic quantum world. There are, however, a number of processes, such as Hawking radiation from the surface of black holes, which can only be understood if quantum mechanics is incorporated in the explanation. But such a successful use of both general relativity and quantum mechanics to explain the same phenomenon is rare. It is only achieved by artificially grafting the quantum rules on top of general relativity in an approximate way. The bottom line is that general relativity and quantum mechanics are incompatible. A symbiosis between these two successful descriptions of reality will only be achieved if they can be merged into one unified scheme; an all-encompassing theory of *quantum gravity*.

Until we find such a theory we will not be able to definitively answer the question of whether or not Hawking's chronology protection conjecture is really a law of nature, thus forbidding time travel. I will end this chapter with a quote from Frank Tipler, the physicist who published the first serious paper on how to build a time machine. Three years after that work, in 1977, he published a longer article in which he examined more carefully the likelihood of his rotating cylinder time machine ever being realized. He ended the article by borrowing a quote from the astronomer Simon Newcomb who had written a number of papers at the turn of the century maintaining the impossibility of heavier-than-air flying machines. Tipler felt that it applied equally well to time machines:

"The demonstration that no possible combination of known substances, known forms of machinery, and known forms of force can be united in a practicable machine by which men shall [travel back in time], seems to the writer as complete as it is possible for the demonstration of any physical fact to be."

I don't need to remind you how Orville and Wilbur Wright proved Newcomb wrong about heavier-than-air flying machines just a few years afterwards.

10
WHAT DO WE KNOW?

"I now believe that if I had asked an even simpler question—such as, What do you mean by mass? or acceleration, which is the scientific equivalent of saying, Can you read?—not more than one in ten of the highly educated would have felt that I was speaking the same language. So the great edifice of modern physics goes up, and the majority of the cleverest people in the western world have about as much insight into it as their neolithic ancestors would have had."

C P Snow, *The Two Cultures*

"Poets say science takes away from the beauty of the stars—mere globs of gas atoms. Nothing is 'mere'. I too see the stars on a desert night, and feel them. But do I see less or more? The vastness of the heavens stretches my imagination—stuck on this carousel my little eye can catch one million year old light; a vast pattern of which I am a part.... What is the pattern, or the meaning, or the why? It does not do harm to the mystery to know a little more about it. For far more marvellous is the truth than any artists of the past imagined it. Why do the poets of the present not speak of it?"

Richard Feynman, *The Feynman Lectures on Physics,*
Volume I

In this final chapter, I wish to take stock of what we believe we know about our Universe and consider how we are likely to make progress.

The mother of all theories

Einstein's geometric theory of gravity (general relativity) and quantum mechanics have been the two greatest achievements of

240

twentieth century physics, towering above everything else that we have learnt about our physical world and beyond, both before and since. Between them, they have cornered the market as far as describing the most fundamental aspects of reality itself goes. The problem, as I mentioned in the last chapter, is that they are just *not* compatible with each other. They rely on very different types of mathematics and have completely separate rules and underlying principles. General relativity breaks down at singularities and closed time loops, while quantum mechanics fails to describe the force of gravity within its framework. So how close are we to a theory of quantum gravity; a 'theory of everything' that will contain within its mathematical structure the rules and principles of both relativity and quantum mechanics? Well, as we see out the twentieth century and begin the twenty first, it may be that we already have such a theory in our grasp.

Einstein completed his general theory of relativity in 1915, and then played a relatively (no pun intended) minor role in the subsequent development of the quantum theory which occupied most of the other leading physicists in the world over the next ten years. But once the ideas and underlying mathematics had been sorted out what else was there to do? Someone of Einstein's genius would not have been content with dotting the 'i's and crossing the 't's. So for the last thirty years of his life he searched, unsuccessfully, for what is called a unified field theory; a theory that would combine general relativity not with quantum mechanics but with the theory of light (or electromagnetism, to be more precise). Einstein tried a number of approaches but never quite cracked it. It is said that papers with his unfinished theory were found on his desk after he died.

The most mathematically elegant, but at the same time most puzzling, of the candidates for a unified theory that Einstein worked on was due to two mathematicians: a Pole, Theodor Kaluza, and a Swede, Oskar Klein. Kaluza did all the ground work and, in 1919, sent a paper to Einstein in which he proposed a way of explaining electromagnetic radiation within the framework of general relativity.

Kaluza showed that what was needed to achieve this was to write his equations not in 4D but 5D spacetime by including an extra space dimension. Although this sounds arbitrary and far removed from what we might consider as reality, it is relatively easy to do mathematically, where we can add as many dimensions as we like. But was this fourth dimension of space that Kaluza was proposing real? We are certainly not aware of it if it is out there. But when he included this extra dimension Kaluza found that light, instead of being an oscillating electromagnetic field through 3D space, was in fact a vibration of this fifth dimension. So there you go. But don't worry, I don't really understand what this means either. All we can say is that it tries to explain the origin of light at a more fundamental, geometrical level in the same way that Einstein had described gravity as a curvature of 4D spacetime. Not only that, but this fifth dimension does not extend in a straight line like the other three dimensions of space but is 'curled up' on itself. A simple way to visualize what this means is to think of 2Dworld. Imagine flat 2D space curled round to make a cylinder. One of the dimensions—the one that points along its length—remains unaltered, whereas the other one has looped around into a circle.

The problem of course was that, despite the elegance of the mathematics of Kaluza's theory, there was not a scrap of experimental evidence whatsoever to suggest that this fifth dimension really existed. Even Einstein, while impressed with the way Kaluza had unified light and gravity, was unwilling to believe in the reality of a fifth dimension. After all, he had been rather reluctant to even take on board the idea of four-dimensional spacetime to begin with. At least the four dimensions (one of time and three of space) were real. The main reason for Einstein's and others' scepticism was because we never see this extra dimension. This question was answered in 1926 when Oskar Klein suggested that the reason it could not be detected was because it was curled up into a circle so tiny that it was billions of times smaller even than an atom. Think again of one of the dimensions of 2Dworld curled round to form the surface of a cylinder. Klein said that the cylinder would be so thin that it would look like a line. That is, 2Dworld would look one-dimensional and we would say that the second dimension was hidden. I am afraid I cannot give you a higher dimensional example than this because, as we saw back in

Chapter 1, we need a third dimension to curve one of 2Dworld's dimensions into. And you thought you had left all that headache-inducing stuff about dimensions behind you.

You aint seen nuthin' yet! Read on.

After many decades in the wilderness, Kaluza–Klein theory made a comeback in the late 1970s. By then the unified theory that the most ambitious theoretical physicists were searching for had to be all singing and all dancing. It was not enough for it to unify gravity and light. By that time it had been established beyond doubt that all phenomena in nature could, at the most fundamental level, be described by four forces. The force of gravity was one and the electromagnetic force another. This latter is the attractive force between electric charges which holds all atoms together by keeping the negatively charged electrons in the grip of the positively charged atomic nucleus. It is also the force of attraction exerted by magnets on each other and on certain metals. I should point out that, despite electric and magnetic forces appearing to be quite separate, this is really only superficial. Michael Faraday had shown in the nineteenth century that they were intimately connected and had their origin in the same electromagnetic force. Almost all phenomena we see around us are due ultimately to one of these two forces: gravity and electromagnetism. We now know there are, in addition to these, two other forces that act only within the tiny confines of the atomic nucleus, but which are just as important as the first two as far as the fundamental laws of nature are concerned.

So the ultimate theory being searched for in the late '70s was one that not only unified gravity with electromagnetism, as Kaluza–Klein theory did, but which also encompassed the two nuclear forces. Such a theory would be called a 'theory of everything', since it would show how all four forces of nature are aspects of just one 'superforce'. The reason Kaluza–Klein theory came back into fashion was because of its clever way of being able to unify forces when higher dimensions of space were included in the equations. Of course with four forces to deal with, instead of just two, more than one extra dimension would be required. Finally, by the mid-1980s a candidate theory was discovered. It was dubbed superstring theory and quickly developed into the most sophisticated, elegant, complicated, powerful and obscure theory ever devised. After all, it was a theory of *ten* dimensions. If correct, it stated that we lived in a ten-dimensional universe. But now all six extra spatial dimensions would be curled up into a tiny high dimensional sphere that we could never detect, leaving just the four dimensions of spacetime. Superstring theory was so named because it suggested that everything is ultimately made of tiny strings which vibrate in ten dimensions. This may sound crazy but it does achieve the unification of general relativity with quantum mechanics which is, after all, the holy grail of physics.

So if superstring theory is the ultimate theory of quantum gravity physicists have been looking for, is the hunt up? And, more importantly for you dear reader, do its equations contain the answer to whether time travel is allowed or not? I am afraid it is still too early to tell yet. Many physicists describe superstrings as a theory of the twenty first century which has been discovered too early: before we have had the chance to develop mathematical tools of the required sophistication. It seems that it is just too hard for anyone to fully comprehend. Its mathematics are beyond the current ability of most, if not all, mathematicians. Not only that, but by the early 1990s there were five different versions of superstring theory and no one knew which was the correct version, or indeed whether there was a unique version.

Then in 1995, a scientist who has been dubbed 'the smartest person on Earth' found an answer, maybe THE answer. His name is Edward Witten and he works at the Institute for Advanced Study in Princeton, New Jersey (which was where Einstein spent his latter years). Together with a colleague, Paul Townsend of Cambridge University, Witten believes he has discovered why there are so many versions of superstring theory. The price that has to be paid is relatively cheap under the circumstances. Witten asks for just one more dimension! With eleven dimensions instead of ten many of the problems of superstring theory go away. Now the tiny strings are replaced by sheets known as membranes and the new theory of Witten and Townsend is called membrane theory, or M-theory for short. However, the 'M' is often taken to stand for

'Magic', 'Mystery' or even 'Mother', since this really would be the mother-of-all-theories.

But could we just keep on going? What if other versions of M-theory were discovered? Maybe adding a twelfth dimension would cure things. In fact, why not just chuck in another handful of dimensions just to be on the safe side! It turns out that this is not possible. There is something mathematically very special about the ten dimensions of superstring theory and the eleven dimensions of M-theory.

Physicists are already making strides in understanding the meaning of M-theory, although I expect it to take several decades before all its magic and mystery is unravelled. One of the main questions to be answered of course is why and how all the extra dimensions get curled round and squeezed down leaving just the four dimensions we see. Current thinking is that this would have happened at the moment of the Big Bang. This implies that there was something *before* the Big Bang. Maybe our three dimensions of space and one of time were part of a much grander ten- or eleven-dimensional universe in which all the forces of nature were unified into one. The Big Bang then caused six or seven dimensions of space to be crushed down to a size that we would never be able to access.

The end of theoretical physics

You might think from the discussion in the previous section that the end of theoretical physics is in sight. Maybe M-theory will answer all our questions, including questions which have until now been thought to be beyond the realm of science. Maybe we lesser mortals should just sit tight while Witten and his colleagues sort out the details of M-theory over the next few years. Then all of fundamental physics would be known. I for one do not subscribe to this view. This is in part due to the fact that I am not an expert in superstring theory or M-theory, and therefore cannot share in the sense of excitement that the practitioners in these fields must feel. But there is another, more justifiable reason for my scepticism.

While many physicists firmly believe we already have a theory of quantum gravity in the shape of the multi-dimensional superstrings or M-theory, there are others who are not so confident. They draw parallels with the state of physics at the end of the last century when it was thought that the end was in sight and that all the laws of nature had been unravelled and understood. Then came the discoveries of x-rays and radioactivity, Max Planck suggested that energy came in discrete packets, or quanta, and Einstein overthrew the Newtonian view of space and time. One hundred years ago no scientists in their wildest dreams could have anticipated what was to happen over the next quarter of a century. So why should we be so confident today? In fact, historians of science point out that back then scientists were probably more justified in thinking that the end of physics was in sight than we are today. Some of the world's leading physicists, such as Roger Penrose and David Deutsch, firmly believe that before quantum mechanics and relativity can be united into a theory of quantum gravity, one or even both of these two theories might need major surgery.

Some ideas in physics stand the test of time and evolve slowly as experimental evidence in their support accumulates. Gradually we understand them more and grow confident that they are a correct description of the physical Universe. Other ideas burst upon the scene suddenly because of an individual moment of genius or a surprising experimental result. But many theories are consigned to the scrap heap when they fail the test of closer scrutiny.

A few successful theories cause a revolution in the way we view the world—known as a paradigm shift in our view. This was the case with Einstein's theory of relativity when he suggested that there was no need for the ether through which light waves would propagate. This led immediately to the conclusion that a beam of light travels at the same speed whether we are moving towards the source of the light or away from it. In turn, this led inevitability to the fact that time runs slower for different observers.

But surely quantum mechanics cannot be wrong, can it? Its predictive power is not in doubt, it's been around for seventy five years and now underpins so much of modern science. Of course, anyone who has learnt something about quantum mechanics will acknowledge just how weird it suggests the microscopic world is, but the standard argument goes like this: The mathematical formalism is right, it is just what the equations *mean* that is not properly understood, and that is a matter of philosophy not physics. The majority of physicists today believe that despite there being a whole host of different interpretations of quantum mechanics on the market, all of which are equally valid, the underlying mathematical framework is correct and it is purely a matter of personal taste which interpretation an individual subscribes to. Whether you believe the Copenhagen interpretation in which nothing exists until it is observed, or the many-worlds interpretation in which the Universe splits into an infinite number of copies, or the Bohmian interpretation in which signals travel faster than light or even, more recently, the transactional interpretation in which signals travel backwards in time, it doesn't matter. No experiment has yet been devised that can discriminate between these rival views. The only thing we are sure about is that quantum mechanics does not have a simple common sense explanation.

In my opinion, to say that the meaning of the mathematical equations that describe reality at its most fundamental level is not important, and that all we should be concerned with are the numbers we obtain by solving these equations, is a cop-out. Over the past ten years or so I, and a growing number of physicists, have become convinced that not all interpretations of quantum mechanics can be right. Nature behaves in a certain way and the fact that we have yet to figure out what is really going on is something we have yet to properly address. For instance, either the Universe splits into many copies of itself or it doesn't. It is tough luck for us if we cannot find out whether this happens or not, but we should not stop trying. We may never succeed in finding what is really going on, but something is going on. I believe that one day we will find out.

As a research student, my hero was the late John Bell, an Irish theoretical physicist and one of the twentieth century's leading experts on quantum mechanics. He was also, as far as I am concerned at any rate, the voice of reason when it came to the interpretation of quantum mechanics. During the 1920s, the two giants of physics, Neils Bohr and Einstein, had a long running debate about the meaning of the then new theory. Einstein argued that quantum mechanics could not be the last word and that there had to be something missing, while Bohr claimed that quantum mechanics told us all we could ever know about nature. Bohr was convinced that physical theories do not describe reality directly but only what we can *know* about reality. His version of quantum mechanics became known as the Copenhagen view since that was where his institute was based. Einstein, on the other hand, felt that a good theory had to be ontological in that it described how reality really is. Quantum mechanics should not be any different. It is generally acknowledged that Bohr won that debate and since then generations of physicists have followed the Copenhagen view.

I am a regular visitor to the Neils Bohr Institute in Copenhagen where much research still goes on today. From the outside it looks like a rather quaint collection of small buildings dwarfed by the nearby large hospital. On the inside, however, it is easy for visitors to become lost in the myriad of tunnels and passages that link the buildings together underground. My real inspiration, however, comes from walking in the park behind the Institute where Bohr and the other giants of early twentieth century physics would spend so much time trying to figure out the strange implications of the new quantum mechanics.

John Bell was of a later generation. I heard him lecture on a number of occasions where he always said he felt that those who adhered to the Copenhagen view were like ostriches with their heads in the sand, not daring to question the deeper meaning of quantum mechanics, but satisfied to blindly follow its rules which worked so well. This troubled Bell since he felt that physics should be about trying to understand the deeper meaning of what was going on in nature.

However, Bell was by no means on the fringes. He has been one of the most respected figures in world physics since the early

1960s and has made some of the most important discoveries in modern physics. I bumped into him for the last time at a meeting of the American Physical Society in Baltimore in 1989, a year before he died. I had attended a fringe meeting on the foundations of quantum mechanics in which the speaker had proposed some new, and clearly dubious, interpretation. I noticed that John Bell was also in the audience. Later that morning, I found myself alone with Bell in a lift going to the cafeteria on the top floor of the conference centre. In order to strike up a conversation with the great man, I asked him what he thought of the last talk.

"Oh, he's clearly wrong" he smiled, "he is obviously not aware of the helium problem".

"Obviously not," I laughed eagerly, wondering what the hell the helium problem might be, but keen to make sure he realized I was in complete agreement with him.

I remember once asking Bell a question after a lecture he gave at Queen Mary College London. He had just argued that he was quite a fan of David Bohm's interpretation of quantum mechanics which describes the whole Universe as being interconnected on the quantum level so that something happening to an atom here on Earth might instantaneously affect another atom in a different galaxy. This type of connectivity between all the particles in the Universe is known as non-locality, or action-at-a-distance, and would require some sort of signalling that must travel faster than light. But surely, I asked Bell, this violates Einstein's special theory of relativity. He replied that he would rather give up special relativity than reality itself, which is literally the price one has to pay if the Copenhagen view is to be believed. You see, according to Bohr nothing even exists in the quantum world until we have measured it and observed it, and since everything is ultimately made up of quantum objects anyway, then nothing (not even the next page of this book) exists until we look at it. Bell maintained that if this were not the case, where would we draw the line between the microscopic world that obeys the quantum rules and the macroscopic world of everyday life?

Astronomy versus astrology

The way many physicists and philosophers are today divided over the meaning of quantum mechanics is rather like the different religious beliefs. Some defend their view passionately and argue that anyone who holds a view other than their own is foolishly wrong. Others are agnostic in that they cannot decide which version of quantum mechanics to 'believe in'. Since one's preferred interpretation is something that cannot be proved, nor can the opposing view be refuted, it becomes a matter of faith. This is not the way science should work, nor in general does it. The following quote is from the physicist Michio Kaku in his book *Hyperspace*:

> *"Some people have accused scientists of creating a new theology based on mathematics; that is, we have rejected the mythology of religion, only to embrace an even stranger religion based on curved spacetime, particle symmetries and cosmic expansion. While priests may chant incantations in Latin that hardly anyone understands, physicists chant arcane equations that even fewer understand. The 'faith' in an all-powerful God is now replaced by 'faith' in quantum mechanics and general relativity."*

So how can non-scientists ever be sure of anything the scientists tell them?

Do not for one minute think that physics is about being open to doubt and uncertainty or that our description of reality is just a matter of personal taste. Today is a Thursday therefore I believe in parallel universes, tomorrow I shall wear my lucky blue socks and so will firmly subscribe to the notion that cosmic strings exist and so on. Science is all about finding the rules that nature follows, discovering a theory and then testing it again and again to see whether it is the correct description of reality. If it fails it is discarded. Many non-scientists often think of us as being too narrow minded and bigoted towards new ideas and possibilities, especially when it comes to things like paranormal phenomena. However, when told that a certain crystal has magical healing powers or that it is able to respond to some sort of psychic energy, a scientist will want to know what form of energy this is, and whether the power it is suppose to have is explainable by the known laws of nature. Can it be replicated? Can it be measured? If it is a new energy or force can its properties be understood? The plain and simple fact is that, so far, and believe me many have searched for over one hundred years now, there has been no scientific evidence whatsoever for any sort of psychic phenomenon. This is not for want of trying or due to a lack of imagination or sufficient open-mindedness on the part of the scientists but rather because all such claims quickly dissolve away in the face of the rigorous demands of scientific inquiry.

Remember that scientists have to be open minded or they would never discover anything new, but they nevertheless take a good deal of convincing when confronted with any new or as yet unexplained phenomenon. A physicist friend of mine, James Christley, once quoted to me the dictum:

"Be open-minded but not so much so that your brain falls out."

This is sound advice. We have come a long way since the age of superstition and magic. Hundreds of years ago, astrology had a strong hold over people. Today most people know that it is nonsense to believe that a distant star, the light from which may have been travelling for thousands of years before it reaches us, could somehow have a real effect on how our daily lives are played out. But in the sixteenth century even astronomers believed in astrology. Another example is the origin of the word 'flu' which is short for the Italian word 'influenza' meaning 'influence' of the planets, because it was believed that they affected our well-being. Do you believe that now or do you accept that there is such a thing as a flu virus?

Science is making advances all the time, and those advances are towards truth and enlightenment. The path is not always straight and we sometimes go up blind alleys, but overall we have made pretty impressive progress. Since I plan to be around for the first half of the twenty first century I hope that during that time we find that the Universe still has many surprises in store for us.

The fascination of science

When some of my colleagues first found out that I was writing a book on wormholes and time machines—remember black holes are respectable—they poured scorn on the project, claiming that it was not *real physics*, that I was selling out to vulgar popularization. This was the stuff of the *X-Files* and had no place in serious science.

It is true that we do not need to grapple with such deep and profound questions as *how* and *why* the Universe came into being to convey the excitement of twentieth century physics. If we care to look around us we see that the whole world is filled with wonder. Why do I not write about that? Why ask questions about what might, or might not, go on in the middle of a black hole when I could be asking such simple questions as 'Why is the sky blue?' 'Why isn't it green or yellow instead?'. Unfortunately, what saddens me is not that most people don't know the answer to this question, but rather that they probably don't care. Anyway, this book has been about sharing my lifetime fascination with the concept of time.

Scientists are a strange breed. No, I don't mean that we are eccentric social misfits, but rather that we have remained childlike in our never-ending desire to want to know 'why'. I find it fantastic that the atoms that make up my body were created inside some distant star billions of years ago; a star that exploded as a supernova, showering the cosmos with its ashes. Some of these ashes then slowly condensed together and heated up again to form a new star, our Sun, and its planets. If you are not awed by this too then we are very different people. But, hey, we can't all be turned on by science; there is too much else going on, and life is short.

I suppose questions about the meaning of time, whether it flows, whether the past and future coexist with the present and whether we will one day be able to visit them are questions which transcend scientific curiosity. In a sense that has made this book easy to write since I have not had to work hard at convincing you that the subject matter is interesting.

Talking of time, it is probably time I ended this book and spent some long overdue quality time with my family. But have I achieved what I set out to do? So time travel to the past may never be possible, wormholes may not exist in our Universe, and there may be nothing on the 'other side' of a black hole. But I wished to get across to non-scientists some of the most profound concepts of space and time, and if they can be made more palatable and interesting by speculating on the possibility of building a time machine then why not?

I hope this book has been entertaining as well as informative. I never set out to write an introductory course in relativity theory, but what I have offered you, I hope, is a glimpse of what modern physics is about and an opportunity to share with me the sheer excitement of contemplating some of the deepest questions of existence. I hope you have enjoyed it.

THE END

BIBLIOGRAPHY

(References marked with a * symbol denote a non-technical book or article that would serve as further reading material.)

* Abbot E 1984 *Flatland: A Romance of Many Dimensions* (New York: New American Library)
* Allen B and Simon J 1992 Time travel on a string *Nature* **357** 19–21
* Asimov I 1972 *Biographical Encyclopaedia of Science and Technology* (London: Pan)
* Barrow J D and Silk J 1983 *The Left Hand of Creation* (London: Unwin)
  Branch D 1998 Density and destiny *Nature* **391** 23–24
  Bucher M A, Goldhaber A S and Turok N 1995 Open universe from inflation *Physical Review* D **52** 3314–3337. Preprint available at xxx.lanl.gov/abs/hep-ph/9411206 on the World Wide Web
* Bucher M A and Spergel D N 1999 Inflation in a low density universe *Scientific American* January, 43–49
  Capek M (ed) 1976 *The Concepts of Space and Time (Boston Studies in the Philosophy of Science)* vol XXII (Dordrecht: Reidel)
  Clark S 1997 *Redshift* (Hatfield: University of Hertfordshire Press)
  Coles P 1998 The end of the old model universe *Nature* **393** 741–744
* Couper H and Henbest N 1998 *To the Ends of the Universe* (London: Dorling Kindersley)
  Curry C 1992 The naturalness of the cosmological constant in the general theory of relativity *Studies in the History and Philosophy of Science* **23** 657–660
  Davies P 1974 *The Physics of Time-Asymmetry* (Surrey University Press–University of California Press)
  Davies P 1983 Inflation and time asymmetry in the Universe *Nature* **301** 398–400
  Davies P (ed) 1989 *The New Physics* (Cambridge: Cambridge University Press)
  Davies P 1995 *About Time* (London: Penguin)

* Deutsch D 1997 *The Fabric of Reality* (London: Penguin)
* Deutsch D and Lockwood M 1994 The quantum physics of time travel *Scientific American* March, 50–56
* Dewdney A K 1984 *The Planiverse* (London: Picador)
  De Witt B S and Graham N (ed) 1973 *The Many-Worlds Interpretation of Quantum Mechanics* (Princeton, NJ: Princeton University Press)
  Droz S, Israel W and Morsink S M 1996 Black holes: the inside story *Physics World* January, 34–37
  Flanagan É É and Wald R M 1996 Does back reaction enforce the averaged null energy condition in semiclassical gravity? *Physical Review D* **54** 6233–6283
  Ford L H and Roman T H 1996 Quantum field theory constrains traversable wormhole geometries *Physical Review D* **53** 5496–5507
  Gold T 1962 The arrow of time *American Journal of Physics* **30** 403–410
  Gott J R 1991 Closed timelike curves produced by pairs of moving cosmic strings: exact solutions *Physical Review Letters* **66** 1126–1129
  Guth A H and Steinhardt P 1984 The inflationary universe *Scientific American* May, 116–120
* Gribbin J 1986 *In Search of the Big Bang* (London: Heinemann)
* Gribbin J 1992 *In Search of the Edge of Time* (London: Penguin; New York: Harmony)
* Gribbin J 1995 *Schrödinger's Kittens* (London: Weidenfeld and Nicolson)
* Gribbin J 1996 *Companion to the Cosmos* (London: Pheonix Giant)
* Halliwell J J 1991 Quantum cosmology and the creation of the universe *Scientific American* December, 28–35
  Halliwell J J, Pérez-Mercader J and Zurek W H (ed) 1994 *Physical Origins of Time Asymmetry* (Cambridge: Cambridge University Press)
  Hartle J B and Hawking S W 1983 The wave function of the universe *Physical Review D* **28** 2960–2975
  Hawking S W 1985 Arrow of time in cosmology *Physical Review D* **32** 2489–2495
* Hawking S W 1988 *A Brief History of Time* (New York: Bantam)
  Hawking S W 1992 Chronology protection conjecture *Physical Review D* **46** 603–611
* Hawking S W 1993 *Black Holes and Baby Universes* (New York: Bantam)
  Hawking S W and Ellis G F R 1973 *The Large Scale Structure of Space-Time* (Cambridge: Cambridge University Press)
* Hay A and Walters P 1997 *Einstein's Mirror* (Cambridge: Cambridge University Press)
  Hod S and Piran T 1998 Mass inflation in dynamical gravitational collapse of a charged scalar field *Physical Review Letters* **81** 1554–1557
* Hogan C J, Kirshner R P and Suntzeff N B 1999 Surveying space-time with supernovae *Scientific American* January, 28–33

* Kaku M 1995 *Hyperspace* (Oxford: Oxford University Press)
* Kaku M 1997 Into the eleventh dimension *New Scientist* 18 January, 32–36
* Kaku M and Thompson J 1995 *Beyond Einstein* (London: Penguin)
* Kaufmann W J 1994 *Universe* (New York: Freeman)
* Krauss L M 1989 *The Fifth Essence: The Search for Dark Matter in the Universe* (New York: Basic)
* Krauss L M 1999 Cosmological antigravity *Scientific American* January, 35–41
  Lachièze-Rey M and Luminet J-P 1995 Cosmic topology *Physics Reports* **254** 135–214. Preprint available at xxx.lanl.gov/abs/gr-qc/9605010 on the World Wide Web
* Luminet J-P, Starkman G D and Weeks J R 1999 Is space finite? *Scientific American* April, 68–75
  Marder L 1971 *Time and the Space-Traveller* (London: George Allen and Unwin)
  Morris M S and Thorne K S 1987 Wormholes in spacetime and their use for interstellar travel: a tool for teaching general relativity *American Journal of Physics* **56** 395–412
  Morris M S, Thorne K S and Yurtsever U 1988 Wormholes, time machines, and the weak energy condition *Physical Review Letters* **61** 1446–1449
* Nahin P J 1993 *Time Machines* (New York: American Institute of Physics)
* Novikov I 1990 *Black Holes and the Universe* (Cambridge: Cambridge University Press)
  Page D N 1983 Inflation does not explain time assymetry *Nature* **304** 39–41
  Pais A 1982 *Subtle is the Lord: The Science and the Life of Albert Einstein* (Oxford: Oxford University Press)
  Peebles P J E 1993 *Principles of Physical Cosmology* (Princeton, NJ: Princeton University Press)
  Peebles P J E 1999 Evolution of the cosmological constant *Nature* **398** 25–26
* Penrose R 1989 *The Emperor's New Mind* (Oxford: Oxford University Press)
  Price H 1989 A point on the arrow of time *Nature* **340** 181–182
  Ray C 1991 *Time, Space and Philosophy* (London: Routledge)
* Rees M 1997 *Before the Beginning: Our Universe and Others* (New York: Simon and Schuster; London: Touchstone)
  Resnick R 1972 *Basic Concepts in Relativity and Early Quantum Theory* (New York: Wiley)
  Riess A G *et al* 1998 Observational evidence from supernovae for an accelerating universe and a cosmological constant *Astronomical*

Journal **116** 1009–1038. Preprint available at xxx.lanl.gov/abs/astro-ph/9805201 on the World Wide Web

Rindler W 1996 *Introduction to Special Relativity* (Oxford: Oxford University Press)

* Sagan C 1985 *Contact* (New York: Simon and Schuster)

Savitt S (ed) 1994 *Time's Arrow Today* (Cambridge: Cambridge University Press)

Simon J Z 1994 The physics of time travel *Physics World* December, 27–33

Smith J H 1965 *Introduction to Special Relativity* (New York: Benjamin)

* Smoot G and Davidson K 1994 *Wrinkles in Time* (New York: Morrow)

Taylor E F and Wheeler J A 1992 *Spacetime Physics* (New York: Freeman)

* Thorne K S 1994 *Black Holes and Time Warps* (New York: Norton; London: Picador)

Tipler F 1974 Rotating cylinders and the possibility of global causality violation *Physical Review* D **9** 2203–2206

Tipler F 1977 Singularities and causality violation *Annals of Physics (New York)* **108** 1–36

Visser M 1993 From wormhole to time machine: remarks on Hawking's chronology protection conjecture *Physical Review* D **47** 554–565

Visser M 1994 van Vleck determinants: traversable wormhole spacetimes *Physical Review* D **49** 3963–3980

Visser M 1996 *Lorentzian Wormholes* (New York: American Institute of Physics)

Visser M 1997 Traversable wormholes: the Roman ring *Physical Review* D **55** 5212–5214

Vollick D N 1997 Maintaining a wormhole with a scalar field *Physical Review* D **56** 4724–4728

Wald R M 1984 *General Relativity* (Chicago: University of Chicago Press)

Weeks J R 1998 Reconstructing the global topology of the Universe from the cosmic microwave background *Classical and Quantum Gravity* **15** 2599–2604. Preprint available at xxx.lanl.gov/abs/astro-ph/9802012 on the World Wide Web

* Weinberg S 1983 *The First Three Minutes* (London: Flamingo)

Weinberg S 1989 The cosmological constant problem *Reviews of Modern Physics* **61** 1–23

* Whitrow G J 1972 *The Nature of Time* (London: Penguin)

* Whitrow G J 1980 *The Natural Philosophy of Time* (Oxford: Clarendon)

* Will C 1986 *Was Einstein Right?* (New York: Basic Books)

Wright A and Wright H 1989 *At the Edge of the Universe* (Chichester: Ellis Harwood)

Wu K K S, Lahav O and Rees M J 1999 The large-scale smoothness of the Universe *Nature* **397** 225–230

* York D 1997 *In Search of Lost Time* (Bristol: Institute of Physics Publishing)

Zlatev I, Wang L and Steinhardt P J 1999 Quintessence, cosmic coincidence, and the cosmological constant *Physical Review Letters* **82** 896-899

* Zwart P J 1976 *About Time* (Amsterdam: North-Holland)

INDEX

expansion of the, 46, 76
  rate of, 66, 71
heat death of the, 58, 137
shape of the, 72, 77
size of the, 77
Visible, 43, 75, 103, 161
universe
  1D, 4
  baby, 214
  clockwork, 115
  closed, 60
  collapsing, 61
  deterministic, 184
  eleven-dimensional, 245
  flat, 59
  infinite, 61
  open, 61, 73
  parallel, 178, 185, 197
  rotating, 220
  static, 47
vacuum fluctuations, 235
van Stockum, W J, 219
Visser, Matt, 196, 210, 237

weak field limit, 217
weightlessness, 27
Weyl, Hermann, 170
Wheeler, John, 87, 199
white dwarf star, 38
white holes, 106
WIMPs (weakly interacting
    massive particles), 70
Witten, Edward, 244
wormhole, 200, 205
  inter-universe, 210
  intra-universe, 210
  Lorentzian, 210
  quantum, 200
  Roman configuration of,
    237

stability of, 210
taxonomy of, 210
traversable, 206

x-ray emissions, 102

Young, Thomas, 80, 143

zero gravity, 28
zero-dimensional, 5